ToF LiDAR for Autonomous Driving

Online at: https://doi.org/10.1088/978-0-7503-3723-6

IOP Series in Advances in Optics, Photonics and Optoelectronics

SERIES EDITOR

Professor Rajpal S Sirohi Consultant Scientist

About the Editor

Rajpal S Sirohi is currently working as a faculty member in the Department of Physics, Alabama A&M University, Huntsville, AL, USA. Prior to this, he was a consultant scientist at the Indian Institute of Science, Bangalore, and before that he was Chair Professor in the Department of Physics, Tezpur University, Assam. During 2000–11 he was an academic administrator, being vice-chancellor to a couple of universities and the director of the Indian Institute of Technology, Delhi. He is the recipient of many international and national awards and the author of more than 400 papers. Dr Sirohi is involved with research concerning optical metrology, optical instrumentation, holography, and speckle phenomena.

About the series

Optics, photonics, and optoelectronics are enabling technologies in many branches of science, engineering, medicine, and agriculture. These technologies have reshaped our outlook and our ways of interacting with each other, and have brought people closer together. They help us to understand many phenomena better and provide deeper insight into the functioning of nature. Further, these technologies themselves are evolving at a rapid rate. Their applications encompass very large spatial scales, from nanometers to the astronomical scale, and a very large temporal range, from picoseconds to billions of years. This series on advances in optics, photonics, and optoelectronics aims to cover topics that are of interest to both academia and industry. Some of the topics to be covered by the books in this series include biophotonics and medical imaging, devices, electromagnetics, fiber optics, information storage, instrumentation, light sources, CCD and CMOS imagers, metamaterials, optical metrology, optical networks, photovoltaics, free-form optics and its evaluation, singular optics, cryptography, and sensors.

About IOP ebooks

The authors are encouraged to take advantage of the features made possible by electronic publication to enhance the reader experience through the use of color, animation, and video and by incorporating supplementary files in their work.

Do you have an idea for a book you'd like to explore?

For further information and details of submitting book proposals, see iopscience.org/books or contact Ashley Gasque at Ashley.gasque@iop.org.

A full list of titles published in this series can be found here: https://iopscience.iop.org/bookListInfo/series-on-advances-in-optics-photonics-and-optoelectronics.

ToF LiDAR for Autonomous Driving

Wei Wei

School of Electronics and Communication Engineering, Guangzhou University, Guangzhou, China

IOP Publishing, Bristol, UK

ISBN 978-0-7503-3723-6 (ebook)
ISBN 978-0-7503-3721-2 (print)
ISBN 978-0-7503-3724-3 (myPrint)
ISBN 978-0-7503-3722-9 (mobi)

DOI 10.1088/978-0-7503-3723-6

Version: 20230701

IOP ebooks

British Library Cataloguing-in-Publication Data: A catalogue record for this book is available from the British Library.

Published by IOP Publishing, wholly owned by The Institute of Physics, London

IOP Publishing, No.2 The Distillery, Glassfields, Avon Street, Bristol, BS2 0GR, UK

US Office: IOP Publishing, Inc., 190 North Independence Mall West, Suite 601, Philadelphia, PA 19106, USA

Contents

Author biography

Wei Wei

Wei Wei received a BS degree in optical information science and technology from Jilin University in 2010, and a PhD in electronic science and technology from the Beijing University of Posts and Telecommunications in 2015. He worked as a Postdoctoral Researcher with University College Cork and Tyndall National Institute from 2015 to 2016, as a Research Engineer with Huawei, Shenzhen, China, from 2016 to 2017, and as a Hong Kong Scholar Researcher with The Hong Kong Polytechnic University from 2018 to 2020. He is currently an Associate Professor with the School of Electronics and Communication Engineering, Guangzhou University. His current research interests include LiDAR, depth estimation, and autonomous driving.

IOP Publishing

ToF LiDAR for Autonomous Driving

Wei Wei

Chapter 1

LiDAR's role in autonomous driving

This Waymo Chrysler Pacifica in Los Altos, 2017 image has been obtained by the author from the Wikimedia website where it was made available by Dllu under a CC BY-SA 4.0 licence. It is included within this article on that basis. It is attributed to Dllu.

Autonomous driving is not totally strange for most of us, it has appeared in many Sci-Fi movies. In recent years, autonomous driving has received a great degree of attention and is entering a pre-industrialization stage with remarkable progress achieved over the past years. Some renowned high-tech companies like Google and Baidu, traditional automakers like Toyota and Volkswagen, and EV-makers like Tesla have all devoted significant effort to autonomous driving. Autonomous driving related startups are becoming very popular for venture capital. This chapter first introduces the concept of autonomous driving and its classification levels. Involved sensors including LiDAR and their comparisons are elaborated. Then, the history of autonomous driving from the perspective of DARPA Grand Challenge, including how LiDAR is involved, is introduced.

doi:10.1088/978-0-7503-3723-6ch1

1.1 Autonomous driving

Autonomous driving normally refers to self-driving vehicles or driverless cars that can have awareness of surrounding conditions, control and planning by computers and have little or entirely no intervention of a human driver [1]. It is important to be noted that *autonomous* here is not exactly the same as *automated*. An automated vehicle like a subway moves along trails, slows down when getting close to a station platform and speeds up as it leaves a station platform according to set programs, whereas autonomous means self-governing, which needs to perceive, make responses and decisions to unexpected objects appearing on roads [2]. Autonomous control implies satisfactory performance under significant uncertainties in the environment, and the ability to compensate for system failures without external intervention.

To perceive surroundings, an autonomous car combines various vehicle-mounted sensors, such as RGB camera, infrared camera, radar, LiDAR, ultrasonic radar, GPS, odometry and inertial measurement units (IMU). The sensing information is fused and processed by advanced algorithms and artificial intelligence (AI) neural networks deployed in the vehicle-mounted on-board computer to recognize obstacles and traffic signs, identify navigation paths and then interpreted by control systems to make right response. Radar sensors monitor the position of surrounding vehicles. RGB cameras recognize traffic lights, road signs, vehicles and pedestrians. LiDAR sensors bounce light pulses off surroundings to measure distances, detect road edges, and identify lane markings. Ultrasonic sensors detect nearby obstacles and other vehicles during parking.

The Society of Automotive Engineers (SAE) developed a classification system that defines the degree of driving automation by six levels of driving automation ranging from 0 (fully manual) to 5 (fully autonomous) [3, 4] (figure 1.1). Level 0 refers to a vehicle with no driving automation technology, but does not mean containing zero sensors or any warning assistance. The human driver is fully in

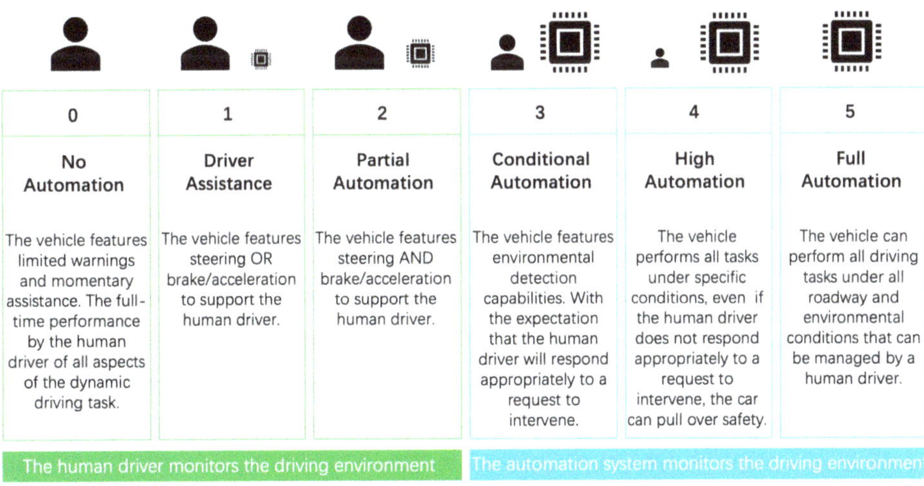

Figure 1.1. SAE driving automation levels.

charge of driving including steering, accelerating, braking, etc, and the vehicle provides limited warnings and momentary assistance. At Level 0, driver support systems such as stability control, forward-collision warning, automatic emergency braking, blind-spot warning and lane-keeping assistance may temporarily intervene when a human driver is driving. These technologies are regarded as Level 0, since they do not take over the control but provide limited warnings and momentary assistance in specific situations. Level 1 is the basic and lowest level of driving automation, providing steering or brake/acceleration assistance to support the human driver. At Level 1, the automation systems start to take control of the vehicle in specific situations, but do not fully take over, the human driver is still responsible for driving the vehicle and must be prepared to take control in any situation. An example of Level 1 automation is adaptive cruise control (ACC), which maintains a safe following distance behind the next vehicle by controlling acceleration and braking without the intervention of the human driver, typically in highway driving. Although the human driver can take their feet off the pedals, steering must be under control by the human driver. Level 2 driving automation applies to vehicles with advanced driving assistance systems (ADAS) that can match steering (lateral control) with acceleration and braking (longitudinal control) in specific scenarios. While the vehicle systems are essentially driving and taking control of primary driving tasks, the human driver is still required to remain alert and engaged, monitor the vehicle and be ready to step in at any time if needed. During Level 2 driving, contact between hand and wheel is generally mandatory to confirm that the human driver is ready to intervene, the eyes of the human driver might be monitored by RGB, infrared or time-of-flight (ToF) cameras to confirm that the human driver is paying attention to traffic. The leap from Level 2 to Level 3 automation is significant and substantial from a technological perspective, the latter is generally considered the initial entry point into autonomous driving. At Level 3, vehicles have 'environmental detection' capabilities with the help of various driver assistance systems and AI to make decisions based on changing driving situations around the vehicle. Human drivers can safely turn their attention away from driving tasks, but only in specific situations. Conditions could be limited to certain vehicle speeds, road types and weather conditions. The human driver can apply their focus to looking at a phone or newspaper. The vehicle handles situations that call for an immediate response, like emergency braking. Nevertheless, the human driver still cannot take a nap while the system is driving, and is expected to take over when the system requests to do so. Level 3 conditional automation can be thought to be a co-driver that alerts you in an orderly fashion when it is your turn to drive. *Audi* announced a Level 3 traffic jam assistance technology called Traffic Jam Pilot for its 2019 A8 flagship sedan, which combines a four-beams LiDAR scanner from *Valeo* (a French global automotive supplier headquartered in France, supplying a wide range of products to automakers and the aftermarket) with advanced sensor fusion and processing power. It might be the world's first mass-production Level 3 vehicle. But it never received regulatory approval for the system in Germany and the regulatory process in the US shifted from federal guidance to state-by-state mandates for autonomous vehicles leading to its classification as Level 2. Level 4 is similar to Level 3, but with higher intelligence to handle more situations, no driver attention is ever required for safety, e.g., the driver may safely go to sleep or leave the driver's seat. At this level, the vehicle's autonomous driving system is fully capable of monitoring the driving environment and handling all driving functions in limited spatial areas (geofenced) or under specific circumstances. Outside of these areas or circumstances, such as severe weather, the vehicle requires a human driver in control. If the human driver does not respond, the driving system will secure the vehicle automatically. Level 5 is the highest classification of driving automation, which means a vehicle can drive itself without any human intervention everywhere in all conditions where an experienced human driver can do so. A Level 5 vehicle is neither bound by geofencing nor affected by weather and won't even have a steering wheel or acceleration/braking pedals. The only human involvement will be to set a destination.

The SAE clarifies that Levels 0–2 are 'driver support features' due to the driver's heavy involvement with the vehicle operation, while Levels 3–5 are 'automated driving features' with more degrees towards autonomous. From Levels 0 to 5, not only the planning and controlling system becomes advanced, the capability of environment monitoring is stronger by mounting more sensors and designing more dedicated algorithms. From Level 3, LiDAR starts to appear on the vehicle to measure the distance of surroundings and further generate point cloud data for advanced algorithms.

LiDAR, camera, millimeter-wave radar and ultrasonic sensor are four typical environmental perception sensors mounted in modern autonomous vehicles, as illustrated in figure 1.2. There are many pivotal driving factors in the selection of environmental perception sensors: cost, detection range, field of view, resolution, depth information, algorithm complexity, robustness, etc. Ultrasonic sensors are the most cost-effective sensors due to their mature technology and wide use in mass-production vehicles. They were mounted on vehicles much earlier than other environmental perception sensors, and were the only environmental perception sensors on vehicles for a period before. As part of the ADAS, ultrasonic sensors have been primarily used for parking guidance and blind-spot detection. As an active sensor that adopts acoustic wave working in ~kHz, ultrasonic sensors can operate in all weathers, day and night. They are perfectly complementary to the other sensors because cameras cannot measure distance and the other two sensors aim to 'see' farther, leaving blind areas in the short distance which can be compensated by ultrasonic sensors. However, in October 2022, the EV-maker Tesla announced removing 12 ultrasonic sensors from Model 3 and Model Y vehicles to save costs, relying on its vision-only approach. Radar that adopts an electromagnetic wave working in ~GHz is used to determine the location, angle and velocity of objects. It can see farthest among all the active sensors. Since it is not as sensitive to, for example dirt, and it does not have any mechanical moving parts, a radar can operate in more varied conditions and environments than a LiDAR. It holds the merits of longer distance detection, speed measurement and stronger penetration in dust over the other three sensors. But it is has poor angular resolution due to the divergence of radio waves. Its resolution does not support distinguishing between different objects.

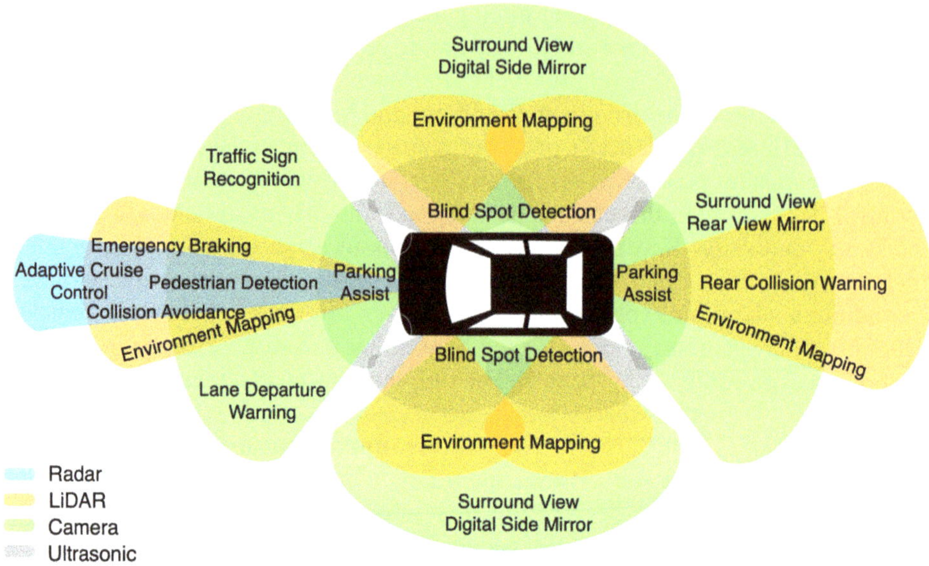

Figure 1.2. Sensor technology in autonomous vehicles. This Car icon top image in the center of the figure has been obtained and reproduced by the author from the Wikimedia website where it was made available by Hedaja under a CC BY-SA 4.0 licence. It is included within this article on that basis. It is attributed to Hedaja.

Currently, its primary application is the ACC to automatically adjust the vehicle's speed to maintain a safe distance from vehicles ahead. Cameras, that use the reflected light from objects to 'see' the surrounding environment similar to our eyes, are the only one passive sensor among all the environmental perception sensors. They are now widely used in vehicles with no autonomous driving systems for parking assistance. Cameras are a primary sensor and indispensable for advanced autonomous vehicles. The cost of a camera is much lower than for radar and LiDAR, and not much higher than for an ultrasonic sensor. But the process of images requires complicated algorithms and high computational resources. Cameras offer much higher resolution than other sensors, making them good at high resolution tasks like classification and scene understanding. They are the only sensor to distinguish colors and accomplish tasks that require color perception like traffic lights or sign recognition. Their disadvantages are that they do not perform well in different weather or lighting conditions; even the night-vision and high dynamic range (HDR)-enabled cameras sometimes struggle with nighttime tasks. Also, they do not offer depth information. Although stereo vision provides depth information computed from a pair of images, the results are estimated via complicated algorithms and not as reliable as active sensors. Detailed comparisons are presented in table 1.1.

Google is a big fan of LiDAR, and mounted a 360° *Velodyne* (a Silicon Valley-based lidar technology company, headquartered in San Jose, California) LiDAR on the roof of its first autonomous car prototype in 2012. After that, the way to measure distance using LiDAR has been adopted by most autonomous driving companies

Table 1.1. Overview of environmental perception sensors on autonomous vehicles.

Sensor type	LiDAR	Camera	MW Radar	Ultrasonic sensor
Max. distance	200 m	200 m	>200 m	<10 m
Frequency	~200 to ~300 THz	Passive	24 GHz, 77 GHz to 81 GHz	>20 kHz
Object detection	✓	✓	✓	✓
Object recognition	✓	✓	×	×
Text recognition	×	✓	×	×
Pedestrian detection	✓	✓	✓	✓
Precision	High	Low	Medium	High is short distance
Work at night	✓	× (✓ for infrared)	✓	✓
Work in bad weather	×	×	✓	✓
Algorithm complexity	Low	High	Low	Low
Cost	High	Low	Medium	Low
Typical applications	3D depth mapping, distance measurement, object detection	Objection detection, traffic sign recognition, trajectory prediction, night vision (infrared)	Collision detection/ avoidance, distance and speed measurement	Parking assist, object detection in short range

and become a consensus for high-level driving automation. But Tesla, who insists on pure passive vision by cameras, says no to LiDAR due to its huge size and cost. Each of these technical routes has its own advantages and disadvantages. They believe that autonomous driving could be realized by cameras, complex neural networks and powerful computing. At this moment, we do not have enough evidence to make a judgement on the two different technical routes. In this book, we only focus on the LiDAR route and make a comprehensive review on LiDAR for autonomous driving. Currently (by Q1, 2023), most autonomous driving startups and manufacturers are still at the phase of Level 2–4, meaning that testing vehicles can move only in strict areas. Although some players announced that they have realized Level 4 autonomous driving or launched new type vehicles integrated with Level 4 autonomous driving systems, a gap still exists between the practical performance and the real Level 4 automation. There are lots of problems to be solved to achieve Level 4 and Level 5 automation, such as technique, cost, regulatory issues and market.

1.2 LiDAR in autonomous driving

Sensors are the interface between the physical world and the digital world, which capture the information of environments and output digitalized data. The data are then processed by perception algorithms to build a perception system for autonomous vehicles to recognize and navigate. An advanced perception system consists of kinds of passive and active sensors. As a representative of active sensors, LiDAR becomes a core sensor for its advantages of actively acquiring abundant 3D position information. Compared to stereo cameras and radar, LiDAR is capable of obtaining much more accurate distance information than stereo cameras without complex algorithms and heavy computing, and higher resolution of distance maps than radar. LiDAR is short for light detection and ranging, which measures distance using the ToF method. The ToF distance measuring technique counts the time interval between the emitted and returned laser pulse along the same path from LiDAR to an object [5]. It was originally utilized in military, remote sensing and surveying applications. The use of LiDAR in autonomous driving can be traced back to 1989; researchers at Carnegie Mellon University's Navlab first used a laser scanner to help the vehicle detect obstacles and determine their range, but the laser scanner was not the primary sensor. In the 2000s, LiDAR found its way into autonomous driving thanks to the DARPA Grand Challenge. The first competition of the DARPA Grand Challenge was held in March 2004, along a 150-mile route in the Mojave Desert region [6, 7]. In this race, there are no drivers, no human behind the wheel and no remote control. It is an autonomous driving race, the first race ever where the machine makes all the decisions. 107 teams registered and 15 raced, but none of the robot vehicles finished the competition. The Red Team from Carnegie Mellon University with its car *Sandstorm* traveled the farthest distance of 7.32 miles (figure 1.3). They chose a combination of several long-range and short-range sensors to make a detection coverage of the areas. Three fixed 2D LiDAR sensors from *SICK AG* (a global manufacturer of sensors and solutions for industrial applications

Figure 1.3. Car *Sandstorm* prior to the 2004 DARPA Grand Challenge. This Sandstorm frontal (DARPA 2004) image has been obtained by the author from the Wikimedia website where it was made available by Rupert Scammel under a CC BY-SA 1.0 licence. It is included within this article on that basis. It is attributed to Rupert Scammel.

based in Germany) operated to characterize the terrain with an overlapping field of view to provide redundancy [8]. One steerable LiDAR sensor from *Riegl* (a laser measurement company based in Horn, Austria, which designs and manufactures precision laser scanners and companion software) with a gimbal on the roof of the car was used to detect obvious obstacles at long ranges. A 2D LiDAR is also called a line scanning LiDAR in contrast to a range finder that measures distance between two points [9].

After the first appearance, LiDAR became popular among teams and then was made famous by *Stanley* in the 2005 DARPA Grand Challenge and then by *Boss* and *Junior* in the 2007 DAPRA Urban Challenge. The 2005 DARPA Grand Challenge was held in the same place with 195 teams registered and 23 raced. This time, five teams finished and car *Stanley* from Stanford University finished the route ahead of all other vehicles and won the competition [10]. To perceive the environment and navigate, all the robot vehicles finishing the route were mounted with LiDAR sensors. The winner *Stanley* (figure 1.4) utilized five roof-mounted *SICK AG* 2D LiDAR sensors, of which beams were pointed forward along the direction of the vehicle with slightly different tilted angles downward to obtain more coverage and build a 3D map of the environment [11]. Thus, the cross-sections of the approaching terrain at different ranges out to 25 m in front of the vehicle could be measured. Behind the five teams, *SICK AG* was the biggest winner and boosted the progress of autonomous driving. From the transition of the configuration from three fixed

Figure 1.4. Car *Stanley* in the DARPA Grand Challenge 2005. This The winner, 2005 DARPA Grand Challenge—Stanford's Stanley image has been obtained by the author from the Wikimedia website https://commons.wikimedia.org/wiki/File:The_winner,_2005_DARPA_Grand_Challenge_-Stanford%27s_Stanley.tiff, where it is stated to have been released into the public domain. It is included within this article on that basis.

LiDAR sensors plus one steerable LiDAR sensor to five LiDAR sensors with different angles, we can find that to achieve better perception of the environment, an autonomous vehicle would need more beams covering more horizontal and vertical scopes and longer distances.

Two years later, with the DARPA Urban Challenge held in 2007, vehicles emerged that could navigate in traffic in a mock urban environment. This required teams to build an autonomous vehicle capable of driving in traffic, performing complex maneuvers such as merging, passing, parking and negotiating intersections. This event was truly groundbreaking, as the first-time autonomous vehicles were capable of interacting with both manned and unmanned vehicle traffic in an urban environment. DARPA admitted 11 vehicles to the final event, of which car *Boss* (figure 1.5) from Carnegie Mellon University and car *Junior* from Stanford University won first and second place, respectively. *Velodyne's* HDL-64E LiDAR, providing high-definition 64 beams, was mounted atop five of the six vehicles that finished the competition. The winner Boss utilized a combination of short-range, and long-range LiDAR sensors [12]. To obtain high-resolution and 360° surrounding distance information of the environment, one roof-mounted 64-beam high-definition LiDAR sensor from *Velodyne* and six 2D LiDAR sensors from *SICK AG* were on the roof and around the body of the vehicle. The detecting distance of the above LiDAR sensors is about 70–80 m. Two LiDAR sensors with 150 m detecting range from *Continental AG* (a German multinational automotive parts manufacturing company) and two LiDAR sensors with 300 m detecting range and

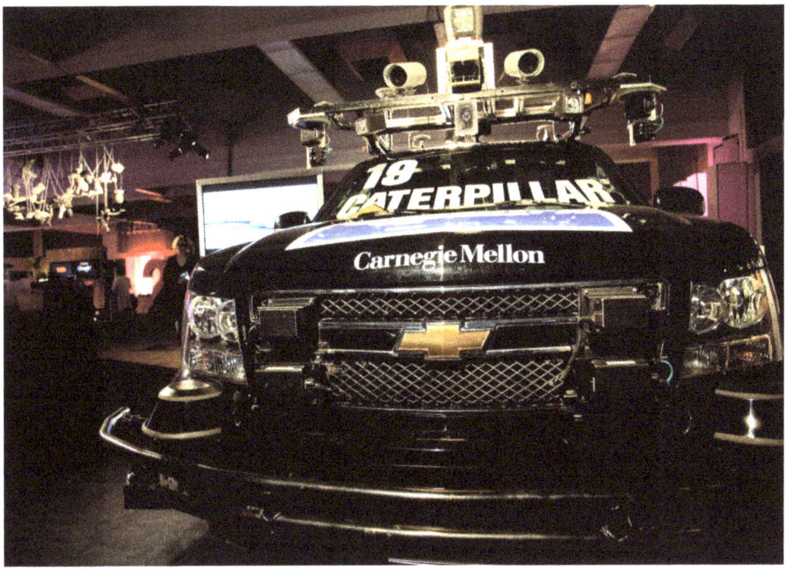

Figure 1.5. Car *Boss* in the exhibition. This CMU carbot (2328347458) image has been obtained by the author from the Wikimedia website https://commons.wikimedia.org/wiki/File:CMU_carbot_(2328347458).jpg where it was made available by Steve Jurveston from Los Altos, USA under a CC BY 2.0 licence. It is included within this article on that basis. It is attributed to Steve Jurveston from Los Altos, USA.

more than 240° horizontal view angles from *IBEO* (a German automotive systems company providing LiDAR sensors, algorithms and software tools) helped *Boss* to 'see' farther and left more time to respond to objects appearing on the road. Similar to *Boss*, *Junior* adopted a combination of LiDAR sensors from *Velodyne*, *SICK AG*, *Riegl* and *IBEO* [13]. The *Velodyne* 64-beam LiDAR sensors mounted on the roof of *Boss* and *Junior* spin to scan all 360° of the surrounding environment. The 64-beam definition can help the vehicle to recognize and distinguish vehicles and pedestrians using advanced algorithms. Since then, LiDAR has occupied a primary position in autonomous vehicles, and a combination of LiDAR, radar and cameras together with deep learning algorithms is the most likely solution of a vast majority of cases.

References

[1] https://en.wikipedia.org/wiki/Self-driving_car (accessed in May 2020)

[2] https://synopsys.com/automotive/what-is-autonomous-car.html (accessed in May 2020)

[3] https://jdpower.com/cars/shopping-guides/levels-of-autonomous-driving-explained (accessed in May 2020)

[4] https://aptiv.com/en/insights/article/what-are-the-levels-of-automated-driving (accessed in May 2020)

[5] Amann M, Bosch T and Lescure M *et al* 2001 Laser ranging: a critical review of usual techniques for distance measurement *Opt. Eng.* **40** 10–9

[6] https://en.wikipedia.org/wiki/DARPA_Grand_Challenge (accessed in May 2020)

[7] Behringer R, Sundareswaran S, Gregory B, Elsley R, Addison B, Guthmiller W, Daily R and Bevly D 2004 The DARPA grand challenge—development of an autonomous vehicle *IEEE Intell. Veh. Symp. 2004 (Parma, Italy, 14–17 June 2004)* 8160812

[8] https://en.wikipedia.org/wiki/Sandstorm_(vehicle) (accessed in May 2020)

[9] Urmson C, Ragusa C and Ray D *et al* 2006 A robust approach to high-speed navigation for unrehearsed desert terrain *J. Field Robot.* **23** 467–508

[10] https://en.wikipedia.org/wiki/Stanley_(vehicle) (accessed in May 2020)

[11] Thrun S, Montemerlo M and Dahlkamp H *et al* 2006 Stanley: the robot that won the DARPA grand challenge *J. Field Robot* **23** 661–92

[12] Urmson C, Anhalt J and Bagnell D *et al* 2008 Autonomous driving in urban environments: Boss and the urban challenge *J. Field Robot* **25** 425–66

[13] Montemerlo M, Becher J, Bhat S and Dahlkamp H *et al* 2008 Junior: the Stanford entry in the urban challenge *J. Field Robot.* **25** 569–97

Chapter 2

Fundamentals of ToF LiDAR

This HELLAS Helicopter Laser Radar image has been obtained by the author from the Wikimedia website where it was made available by Waerfelu under a CC BY-SA 3.0 licence. It is included within this article on that basis. It is attributed to Waerfelu.

LiDAR attracts much attention from researchers and engineers in research institutes and industry, and has been a well-known measurement technique. As a key sensor in autonomous driving, LiDAR is pushing forward the progress of autonomous driving towards practical implementation on roads. The diversity of state-of-the-art approaches brings uncertainty to the trend of dominant technical solutions. In this chapter, we present an overview of the ranging principle including direct time-of-flight (ToF), indirect ToF and frequency-modulated continuous wave (FMCW), and the structure of LiDAR. This chapter first gives a brief introduction on ranging methods of direct ToF, indirect ToF and FMCW, and goes deep into the direct ToF method. Then, structure of LiDAR including diverse scanning schemes is introduced. Finally, principles of indirect ToF and FMCW are elaborated.

2.1 Principle of ToF ranging

The main function of a LiDAR is to measure the distance to an object by laser beams, normally referred to as 'ranging' [1]. A single-point LiDAR is also called a rangefinder that measures the distance point to point. By combining with a scanning system, a cluster of point cloud carrying distance information of the surrounding environment can be obtained to generate 3D depth maps. The manner of distance measurement for LiDAR depends on the type of signal modulation used in the laser beams. The most widely method adopted by LiDAR companies is the direct ToF approach, which modulates laser beams in time domain. The ToF approach generally indicates direct ToF, which emits laser pulses to the target and measures the distance by counting the round-trip time of laser pulses. The indirect ToF approach is based on the amplitude modulation of a continuous wave (AMCW). It measures distance by comparing the phase shift between the original and returned signals. By modulating the frequency of a continuous wave, distance and velocity can be measured via the Doppler effect. This frequency-modulated continuous wave (FMCW) approach is also known as coherent detection, which is similar to the technique commonly used in long-haul optical communications. This book focuses on the direct ToF approach, as it is a popular ranging technique adopted by major LiDAR companies. Also, brief overviews on AMCW and FMCW LiDAR will be presented in sections 2.2 and 2.3.

As in the ToF principle illuminated in figure 2.1, distance between the transmitter and target is determined by multiplying the speed of light in air by the time a laser pulse takes to make a round trip from the transmitter to the target [2]. Since a laser pulse travels to the target forth and back, the measured time is a round-trip time, representing twice the distance to the target. So, the actual distance is halved as:

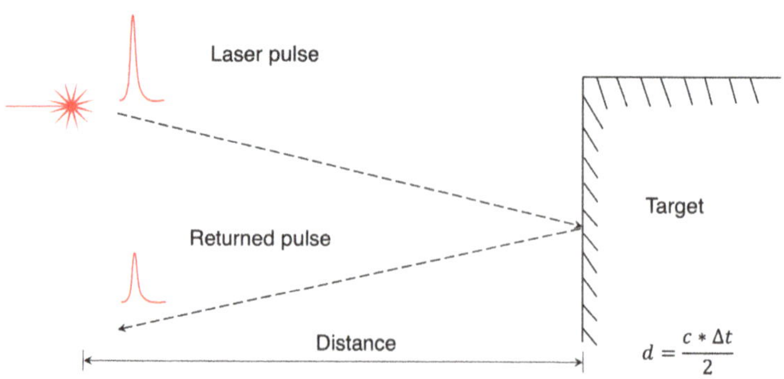

Figure 2.1. Schematic diagram of ToF ranging principle.

$$d = \frac{c \times \Delta t}{2} \tag{2.1}$$

where, d is the distance to the target, c is the speed of light in free space (300 000 km s^{-1}) and Δt is the round-trip time. Since the speed of light is constant in an unchanged medium, the distance is directly proportional to the measured round-trip time. Thus, the ranging resolution is determined by the accuracy of time discriminating and counting electronics, which will by elaborated in chapter 4. Therefore, a cm-level resolution in depth requires the measured time interval around 0.06–0.6 ns. To get better discrimination, pulses need to be as short as possible (normally several ns to 100 ns) with fast rise and fall times. Due to the limitation of jitter and noise in time discriminating and counting electronics, time resolution is challenging to go beyond 0.01 ns. If sacrificing the key parameters like frame rate or angular resolution, time resolution can be improved using statistics, but several pulses are required to measure one point.

An object can be assumed to be as an approximation to Lambertian scatter (or Lambertian diffuser), on which light is scattered, there can be different angular distributions of the scattered light [3]. For an object at distance D, the amount of power received by the photodetector from a pulsed laser transmitter can be approximately modeled as:

$$P_R = P_T \times \frac{A}{2\pi D^2} \Gamma \eta_T \eta_R \tag{2.2}$$

where P_R is the received power, P_T is the peak power of a laser pulse, A is the area of the opto-electronic sensing element, D is the distance of the target, Γ is the reflectivity of the target, η_T is the efficiency of the transmitter and η_R is the efficiency of the receiver. Thus, the received power is a function of square of the target distance. For typical detecting distance of 100 m, the received power can be as low as several hundred nW. Moreover, due to external light sources caused by mutual interference, the signal-to-noise ratio (SNR) is usually not high enough. Attempts to increase SNR usually require delicate design of receiver electronics and high transmitter power. However, for the consideration of eye-safety, simply increasing transmitter power is limited by the eye-safety standard of IEC 60825 [4]. Shorter pulse width of laser can achieve higher peak power while maintaining the eye-safety exposure. However, this requires wide bandwidth of receiver electronics, bringing on larger noise and jitter and increasing the complexity of circuit design.

2.2 LiDAR structure

A LiDAR is a complicated system that consists of diverse functional modules. A typical ToF LiDAR consists of three modules, transmitter, receiver (including timing discriminator), and scanner, as illustrated in figure 2.2. LiDAR sensors from different companies have various structures, but all have the above three functional modules. A transmitter is to provide laser pulses that meet certain requirements depending on application needs, e.g. wavelength, linewidth, frequency accuracy, pulse width, pulse energy, repetition rate, etc. Normally, the transmitter consists of a

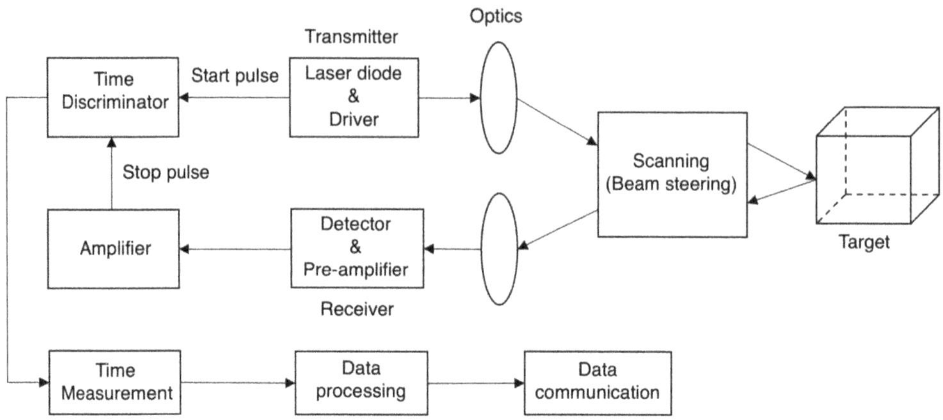

Figure 2.2. ToF LiDAR structure.

laser diode, a driver circuit and optical lenses. Certainly, the laser diode can be replaced by another laser like a fiber laser or solid-state laser. However, a semiconductor with very low cost and tiny dimensions is of greater advantage. The operating wavelength of laser used by ToF LiDAR sensors falls in the infrared region, which is invisible to the naked eye. A laser diode is a current-driven component, in which photons are generated by the transition of electrons. To emit a pulse laser with the required short pulse width and high peak power, the laser diode needs a driver offering transient strong current within a short time. Thus, the performance of the driven circuit directly decides the quality of laser pulse. The laser beam emitted from the laser diode is extremely diverged. The diverged beam leads to the laser spot hitting the object within a large area, causing an adverse effect on the spatial resolution and detecting distance. So, optical lenses with matched parameters are required to collimate the diverged beam to reduce energy loss. A receiver is to collect the detect returned optical signals while compressing the background noise. It usually consists of a photodetector, an amplifier, a timing discriminator, a time measurement unit and focus lenses. A photodetector is a semiconductor chip with dimensions around several hundred nanometers, with which it is difficult to detect weak returned laser pulses due to its tiny sensitive area and the disturbance of ambient light. So, focus lenses with larger aperture and smaller field of view (FOV) are required to collect more laser pulses and suppress ambient light. To further increase SNR, an optical filter is also required to make the photodetector only detect light with a specified wavelength. After a long travel distance, the power of laser pulses detected by the photodetector is extremely weak, and so is the converted photocurrent by the photodetector. The signal manipulated and processed by chips is generally the voltage. So, the photocurrent needs to be converted to voltage by the transimpedance amplifier (TIA), which is also known as the pre-amplifier. The amplitude of voltage signal converted by the TIA is normally not high enough to be processed by the time discriminator. A second amplifier is used to further amplify the voltage signal. Finally, a time discriminator and time measurement unit are adopted to discriminate the time of the returned pulse signals and measure the travelling duration time, which directly determines the ranging accuracy of LiDAR [5]. To get a large FOV and cover more detecting areas, a scanning unit is required to steer the illumination direction of beams. The FOV, frame rate and spatial resolution are determined by the structure of the scanning unit. There exist two kinds of optical structure in LiDAR, coaxial and biaxial arrangements. In a coaxial system, the axis of the laser beam is coincident with the axis of the receiver optics. While in a biaxial system, the laser beam only enters the FOV of the receiver optics beyond some predetermined range. The biaxial arrangement avoids near-field backscattered radiation saturating the photodetector. Additional methods of either gating of the photodetector or use of a fast shutter or chopper are required to overcome the near-field backscattering problem in a coaxial system.

As mentioned above, a point-to-point ranging device is normally referred to as a rangefinder. By mounting a horizontal rotating unit, a rangefinder with 2D scanning capability is known as a 2D LiDAR. Furthermore, a 3D LiDAR extended to X–Y scanning can be obtained by adding a vertical rotating unit. According to scanning approaches, LiDAR are usually classified as mechanical, hybrid solid-state and solid-state, as illustrated in figure 2.3. A mechanical LiDAR contains a bulky rotating system, and all the transceiver modules and control boards are mounted on the rotating platform and spin with the rotating system. It was the most popular scanning approach at the early stage of the development of LiDAR. The *Velodyne* HDL-64E LiDAR in the early ages of autonomous driving history was a typical mechanical LiDAR. Intuitively, a mechanical LiDAR rotates in the horizontal dimension and the vertical movement can be realized by tilting laser beams via an embedded nodding mirror. However, to obtain excellent robustness, a commercial product normally uses precisely aligned multiple transceivers to reduce the movable mechanism. Thus, to increase the vertical resolution, the extra transceivers needed will make the mechanical system significantly large and bulky. Hybrid solid-state

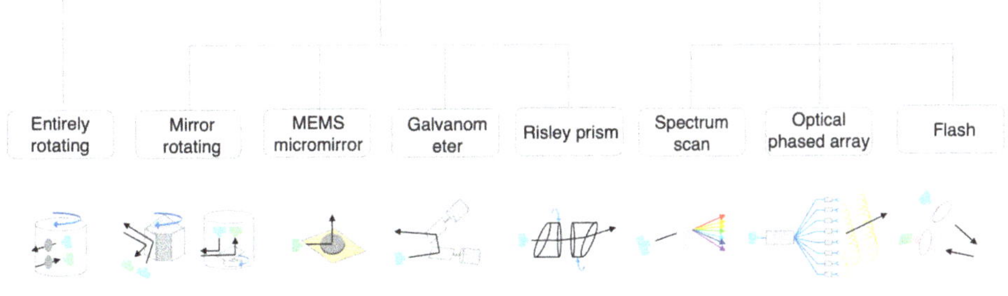

Figure 2.3. Classification of ToF LiDAR.

LiDAR recently became popular due to its compactness and robustness. A hybrid solid-state type also contains a mechanical rotating mechanism. But the transceiver stays fixed and only the mirror rotates to steer beam scanning. The above-mentioned mass-production *Valeo Scala* LiDAR is a hybrid solid-state LiDAR with a rotating mirror. In a mirror rotating LiDAR, the vertical layers can be increased either by using a linear laser diode array, a linear photodiode array or a polygon mirror with different tilted angles. A galvanometric mirror LiDAR based on a galvanometer mechanism uses electromagnetic force to vibrate mirrors along *X*- and *YY*-axis. The rapid vibration steers the beam scanning line by line, bringing high angular resolution and frame rate. MEMS technology allows the fabrication of miniature electro-mechanical devices using silicon fabrication techniques [6]. A MEMS mirror is in essence a mirror embedded in a chip and vibrates via balancing between two opposite forces of electromagnetic and elastic forces. It works either in resonant mode at its characteristic oscillation frequency to achieve a large deflection angle and high operating frequency or in non-resonant mode to achieve controllable programmed scanning trajectory. LiDAR manufacturers like *AEYE* and *Robosense* have released MEMS-based LiDAR products. DJI, a leading manufacturer of drones based in China, released their first LiDAR product that adopts Risley prisms as scanners. The novel design and excellent control on hardware bring the cost of LiDAR down to the range below 1000 US dollars. Although, hybrid solid-state LiDARs are less bulky compared to mechanical ones, autonomous driving manufacturers seek more compact and robust modules with low-cost and easy-to-use merits as CMOS cameras. A solid-state LiDAR that totally removes the rotating parts within scanning systems satisfies the above requirements. Solid-state LiDAR can be divided into two kinds according to whether scanning or not. An optical phased array acquires dense point cloud data still depending on scanning line by line. But, similar to the phased array radar, it contains no moving components and beams are steered by controlling the optical wave-front shape via optical phase modulators. The optical phased array technology requires complicated semiconductor fabrication techniques and controlling algorithms. Quanergy is the first startup that announced releasing high-resolution LiDARs based on optical phased array. But the mass production of optical phased array is so challenging that no robust products with good performance have been released by Quanergy. Dispersion-based scanning adopts the physical principle that a light deflects with varying angles at the interface of two dielectrics with different refractive indices, when its wavelength differs. Scanning is accomplished by varying the emitted wavelength of a wavelength-tunable laser, while a flash LiDAR that behaves as a camera further removes the scanning system. Being spread by an optical diffuser, a single laser illuminates the whole scene at once. Similar to the CMOS camera sensor, a 2D array of photodiodes is adopted to capture all the returned laser signals and generates a dense point cloud. Owing to advanced semiconductor fabrication and packaging techniques, the size and the cost of flash LiDAR can be reduced by a lot. As the pixel inside a photodiode array is more complicated than a CMOS sensor, the semiconductor fabrication limits the spatial resolution of the photodiode array chip. Moreover, the spread illumination lowers the laser power, resulting in limited detecting distance. While increasing emitting power, eye-safety regulations must be complied with.

2.3 AMCW LiDAR

Another ToF approach of ranging is based on AMCW, so the phase shift between the emitted and backscattered waves induced in an intensity-modulated periodic signal are compared enabling one to measure distance. This principle is also known as indirect ToF or i-ToF. As illustrated in figure 2.4, the IR (infrared) emitter, usually a CW laser diode or LED, illuminates periodically intensity-modulated waves with a few tenths of MHz to the object. Due to the flight distance, phase shift $\Delta\varphi$ occurs between the reflected and emitted waves depending on the object distance. By comparing the round-trip phase shift, the distance of the target can be calculated.

Various techniques are used to demodulate the received signal and to extract phase information from it. Generally, the phase shift can be measured and converted to distance using a gate control signal to control the electric charge stored by capacitors, as illustrated in figure 2.5. The emitted wave can be either square or sinusoidal. Square wave makes it easier to calculate phase shift, since the accumulated electric charge is highly linear with time. By controlling the gate signal, the accumulation of electric charge varies with the reflected signal at different distances. Reflected signal at farther distance accumulates more electric charge than that at closer distance. The distance can then be calculated by the phase shift and modulation frequency as [7]

$$d = \frac{c}{2f}\frac{\Delta\varphi}{2\pi} \tag{2.3}$$

where c is the light speed and f is the modulation frequency. In practical applications, one distance measurement is accomplished by accumulating multiple periods to increase accuracy and suppress noise. An accuracy comparable to or better than that of the pulsed technique can be achieved but only at moderate ranges,

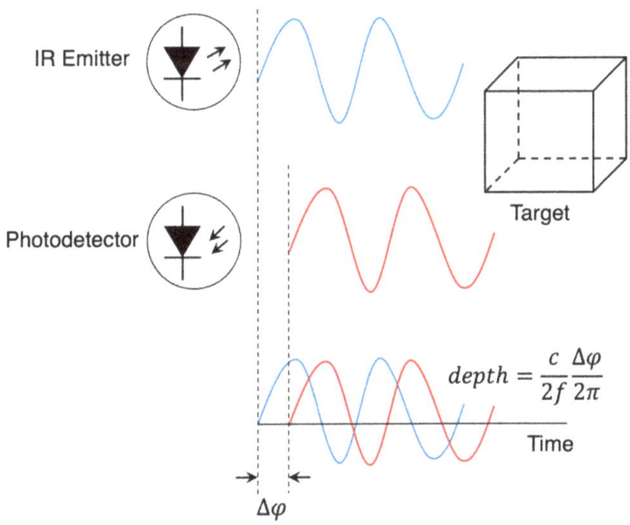

Figure 2.4. Principle of AMCW LiDAR.

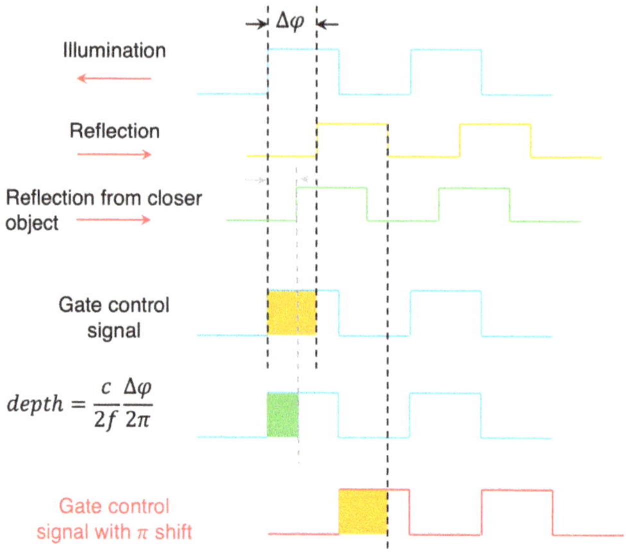

Figure 2.5. Distance calculation by sampling electric charges accumulated using gate control signal.

due to the short ambiguity distance imposed by the 2π ambiguity in frequency modulation. The distance resolution is determined by the modulation frequency f and the demodulation electronics. The quantity $c/2f$ is the maximum distance that can be measured without ambiguity. Larger f brings on shorter unambiguous distance measurements, meaning that the phase value of the reflected signal at different distance values starts to repeat itself after a 2π phase shift [8]. Thus, a significant trade-off appears between the maximum unambiguous distance and resolution of the measurement. The reflected signal arriving at the receiver coming from distant objects is also not as strong as in the pulsed case, as the emission is continuous, which makes the amplitude remain below the eye-safe limit at all times. The rather low emission power compared to daylight limits the measurement precision and applicable scene to mostly indoors, as daylight contaminates the reflected signal. The advantage of this technology is high frame rate due to low computation required, small footprint and relatively low cost.

Sometimes, an additional gate control signal with π phase shift is employed to increase the ranging accuracy and reduce noise by calculating the ratio of two accumulation values. To get higher accuracy, a ToF camera by *Melexis* adopts a quadrature sampling technique [8]. The phase difference is obtained from the relation between four different electric charge values. The four phase control signals have phase delays 90° from each other, determining the collection of electrons from the detected lightwaves. The phase shift is calculated by the four electric charge values as

$$\Delta\varphi = \arctan\left(\frac{Q_4 - Q_3}{Q_1 - Q_2}\right) \tag{2.4}$$

where Q_1 to Q_4 represent the amount of electric charge for the four phase control signals C_1 to C_4 (Q_1 for 0°, Q_2 for 180°, Q_3 for 90°, and Q_4 for 270°). The distance is also calculated as in equation (2.3).

Arrayed photodetectors with capacitors controlled by signals can be integrated into a monolithic chip to reduce cost and footprint. The i-ToF depth sensor illuminates IR waves to the scene, and measures the phase shift of reflected waves at each sensor pixel. Similar to a color camera, an i-ToF camera produces a depth image, each pixel of which encodes the distance to the corresponding point in the scene. Due to the color, reflectivity, geometric structure of target and power consumption limit, the reflected signal suffers from low SNR. The SNR can be increased by collecting multiple periods of modulated signals or binding multiple sensor pixels to calculate a single depth pixel value, which sacrifices frame rate or resolution. Due to its compact size and low cost, an i-ToF camera shows excellent indoor performance in interactive video games.

2.4 FMCW LiDAR

Rather than modulating the amplitude of the signal, the FMCW technique modulates and demodulates the signal in frequency domain, allowing detection by a coherent superposition of the emitted and detected signals [9, 10]. The emitted instantaneous optical frequency is periodically shifted by periodical variation of the driven current of the laser diode (linearly chirped lasers). As illustrated in figure 2.6, the periodically frequency-modulated wave commonly in functions of sawtooth or triangular waves is emitted to the target, while a portion of keeps a reference signal referred to as a local oscillator. In the coherent detection of the FMCW LiDAR, the

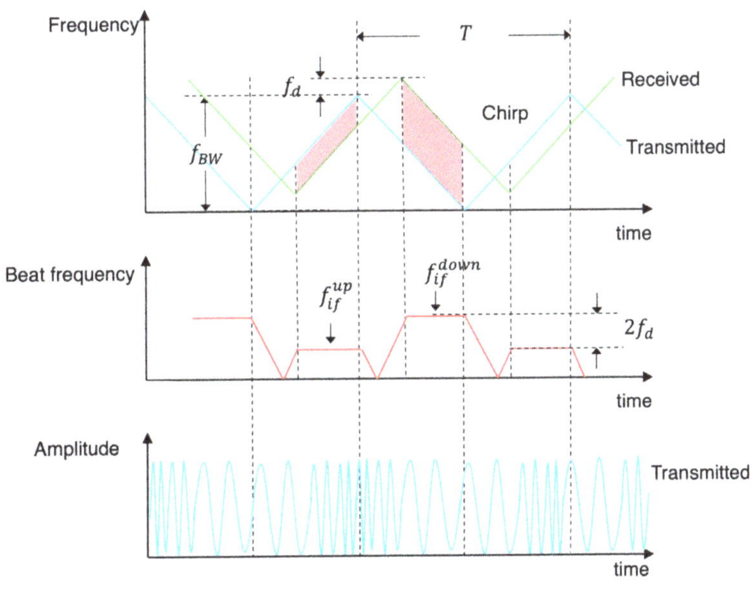

Figure 2.6. Principle of coherent detection.

returned wave with a delay after travelled time arriving at the balanced photodetector is mixed with the reference signal, yielding a time-varying intermediate frequency (red trace). The target distance is directly related to the intermediate frequency as:

$$d = \frac{cTf_{\text{if}}}{4f_{\text{BW}}} \tag{2.5}$$

where T is the waveform period, c is the speed of light, and f_{BW} is the modulation bandwidth of the frequency sweeping.

The FMCW LiDAR has advantages of simultaneously measuring both distance and velocity of the target using the same signal via the Doppler effect. In the case of a moving target, a Doppler frequency shift will be superimposed to the intermediate frequency, that is, an increasing/decreasing change in frequency over the up-ramp/down-ramp waveform [11]. Thus, the intermediate frequency f_{if} in equation (2.5) can be replaced by

$$f_{\text{if}} = \frac{f_{\text{if}}^{\text{up}} + f_{\text{if}}^{\text{down}}}{2} \tag{2.6}$$

By assuming that the Doppler frequency shift is less than the intermediate frequency, the Doppler frequency shift and the velocity of the target can be obtained by

$$f_{\text{d}} = \frac{f_{\text{if}}^{\text{up}} - f_{\text{if}}^{\text{down}}}{2} \tag{2.7}$$

$$v = \frac{\lambda f_{\text{d}}}{2} = \frac{\lambda}{4}(f_{\text{if}}^{\text{up}} - f_{\text{if}}^{\text{down}}) \tag{2.8}$$

where λ is the wavelength of laser.

By exploiting the large frequency bandwidth in the optical domain and coherent detection scheme in the Fourier domain, FMCW is fundamentally different from the aforementioned two ranging approaches. It performs better ranging precision and availability in outdoor environments relative to the AMCW and pulsed ToF approaches. Its ranging precision is related to the total bandwidth of the signal, and millimeter or submillimeter precision can be realized by performing frequency measurements in the kilohertz regime. An FMCW LiDAR continuously illuminates the objects with less emitted power complying with eye-safe requirements. However, the coherent detection scheme has rigid requirements on the quality of the laser source, such as narrow linewidth, good temperature stability and linear chirp. The demodulation complexity is consequently higher than the previous two approaches. Although the pulsed ToF LiDAR still holds a significant position in industry, more and more FMCW LiDARs operating at 1550 nm have emerged.

References

[1] Li Y and Ibanez-Guzman J 2020 Lidar for autonomous driving: the principles, challenges, and trends for automotive lidar and perception systems *IEEE Signal Process. Mag.* **37** 50–61

[2] Royo S and Ballesta-Garcia M 2019 An overview of LiDAR imaging systems for autonomous vehicles *Appl. Sci.* **9** 4093
[3] https://rp-photonics.com/lambertian_emitters_and_scatterers.html (accessed in September 2020)
[4] Safety of Laser Products—All Parts 2021 *International Electrotechnical Commission IE* **60825** 2021
[5] Chevrier M and Campanells G 2017 How to build a LiDAR system with a time-to-digital converter *Analog Appl. J.* **1Q** 1–6
[6] Roriz R, Cabral J and Gomes T 2021 Automotive LiDAR technology: a survey *IEEE Trans. Intell. Trans. Syst.* **23** 6282–97
[7] Time-of-flight Basics 2022 *Melexis: Application Note* 1–14
[8] Hansard M, Lee S, Choi O and Horaud R 2012 *Time of Flight Cameras: Principles, Methods, and Applications* 1st edn (London: Springer) p 95
[9] https://semiengineering.com/automotive-lidar-technologies-battle-it-out/ (accessed in September 2021)
[10] Pierrottet D, Amzajerdian F, Petway L, Barnes B, Lockard G and Rubio M 2008 Linear FMCW laser radar for precision range and vector velocity measurements *MRS Proc.* **1076** 10760406
[11] Agishev R, Gross B, Moshary F, Gilerson A and Ahmed S 2006 Range-resolved pulsed and CWFM lidars: potential capabilities comparison *Appl. Phys.* B **85** 149–62

Chapter 3

Laser source and transmitter

This Kyocera FS-C5200DN—laser unit—board with diode laser-5407 image has been obtained by the author from the Wikimedia website where it was made available by Raimond Spekking under a CC BY-SA 4.0 licence. It is included within this article on that basis. It is attributed to Raimond Spekking.

As an active ranging approach, a transmitter in a LiDAR illuminates specific laser signals with required modulation type, power, wavelength, beam divergence, and repetition rate, directly determining the measuring distance, signal-to-noise ratio (SNR) and accuracy. The core of a transmitter is a laser diode with the capability of

doi:10.1088/978-0-7503-3723-6ch3

converting electrons to photons and emitting laser at near-infrared region. The driver circuits provide required current and control the emission of laser signals. The active detecting property makes LiDAR available in dark environments without the help of any natural or artificial light source. In this chapter, the theory of lasers and semiconductor laser diodes is first introduced. Then, edge emitting laser and driver circuits are elaborated. In addition, examples of driver circuits are given at the end of this chapter.

3.1 Fundamentals of semiconductor lasers

Since its discovery, laser has been demonstrated in solid, liquid, gas and plasma materials [1]. Semiconductor laser diodes are the most widespread used laser source due to their merits of low cost and high compactness. A laser diode, like most other laser sources, consists of an optical gain medium and two reflectors (also called mirrors) forming an optical resonant cavity, as illustrated in figure 3.1. An optical resonant cavity is a key component of all laser sources, which surrounds the gain medium and provides feedback for light waves inside. Photons are generated in the gain medium, where incident radiation is absorbed by electrons.

Electrons in semiconductor material naturally tend to fill up the low energy bands first, while the higher energy bands are basically empty. In figure 3.2, when a gain medium is pumped optically or electrically, electrons inside the medium can be pumped or excited from a lower energy level to an upper energy level. As upper energy levels are generally less stable compared to lower energy levels, the excited electrons are relaxed to lower energy levels by radiating energy in a form of photons. This process is referred to as spontaneous emission, for which the direction and phase of emitted photons are random. The generated photons travel round-trip in the cavity due to the existence of the reflectors. The light inside the cavity reflects multiple times, passing through the gain medium, getting amplified rather than absorbed by stimulating the de-excitation of these electrons along with the generation of additional photons, and producing standing waves at certain

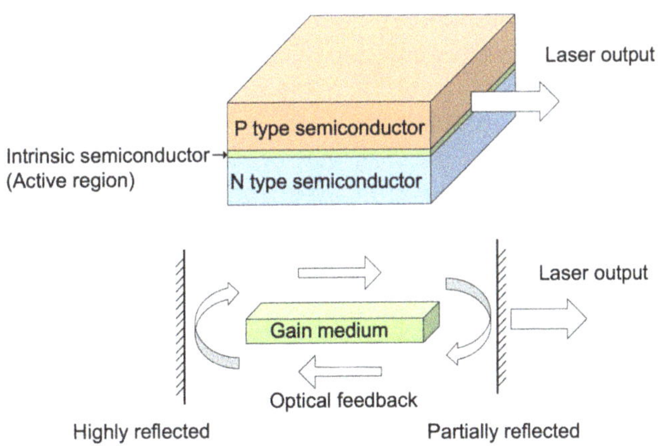

Figure 3.1. Schematic diagram of semiconductor laser diode and resonant cavity.

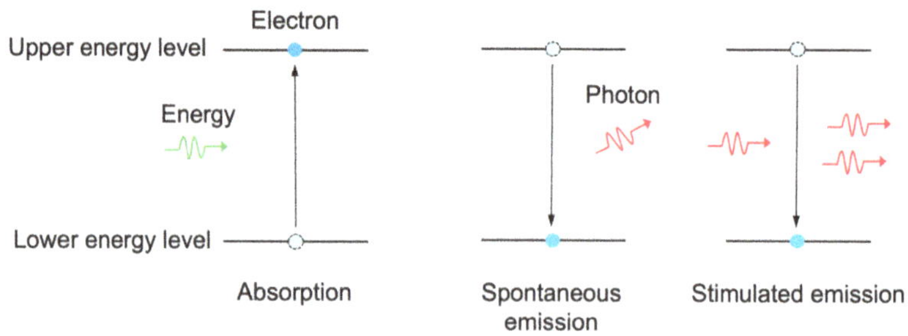

Figure 3.2. Absorption, spontaneous emission and stimulated emission.

resonance frequencies. The produced standing wave patterns are referred to as modes and classified by longitudinal modes and transverse modes. The cavity where standing waves travel is technically known as a *Fabry–Pérot* resonant cavity (figure 3.1) formed by cleaved semiconductor facets and gain medium [2]. Thus, three conditions required for laser oscillation can be summarized as high optical gain, population inversion, and resonator structure. Longitudinal modes differ only in frequency while transverse modes differ not only in frequency but also intensity patterns across the cross-section of beams. The response of the optical reflectors can be tailored to support single mode or few modes operation by incorporating additional optical filtering elements inside the resonant cavity. Normally, a single mode operation is preferred in various applications.

Most semiconductor laser diodes are electrically pumped, where electron–hole pairs are generated by an electrical current in a region where n-doped and p-doped semiconductor materials meet, as illustrated in figure 3.3. Forward biased by electric field, the p–n junction allows recombination of electrons with holes in the intrinsic region. The drop of the electron from an upper energy level to a lower one results in energy radiation in the form of a photon. When the excited electron interacts with another photon inside the cavity, stimulated emission happens. Different from spontaneous emission, both the direction and phase stay the same with the other photon in the process of stimulated emission, which is highly featured in coherent light sources, i.e., laser. Another difference is that the stimulation emission requires a strong pump together with the creation of population inversion. A population inversion represents a sizable population of electrons residing in upper energy levels, and it sets the stage for stimulated emission of multiple photons. This is the precondition for the light amplification that occurs in a laser and makes the light highly coherent. When pumped at a lower intensity, output emission is weak and only spontaneous emission exists in the cavity. If strongly pumped above the threshold and the gain is sufficient to overcome the loss of resonant modes in the cavity, stimulated emission occurs and dominates in the resonant cavity, which makes output emission strong and extremely bright.

The properties of semiconductor materials make the fabrication of laser diodes feasible. Laser diodes almost exclusively use direct bandgap semiconductor materials. Different from the traditional non-light-emitting semiconductor junction diode (indirect bandgap semiconductor material, i.e., silicon or germanium, where most

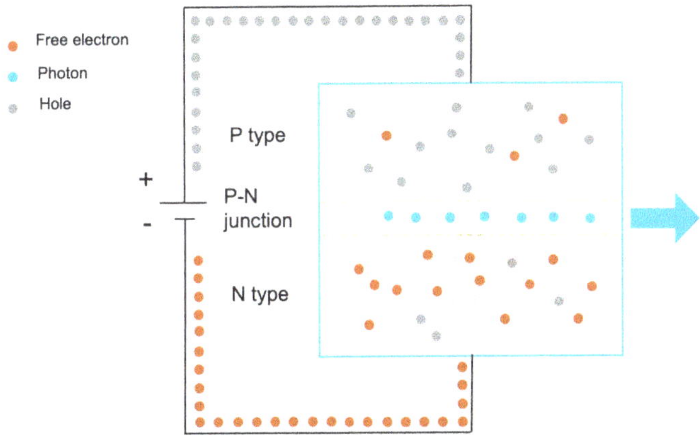

Figure 3.3. Electrically pumped laser diode.

energy is converted to heat), the crystal momentum of electrons and holes in the direct bandgap semiconductor material is the same in both the conduction band and the valence band enabling direct emission of a photon from an electron. As a typical direct bandgap semiconductor material, III–V compound semiconductors including GaAs, InAs and InP are widely used to fabricate laser diodes and other optoelectronic components like photodiodes and modulators. A III–V compound semiconductor is an alloy consisting of elements from columns III and V in the periodic table. Depending on the number of elements involved, these III–V compound semiconductors can be either binary, ternary or quaternary, covering wavelength range of laser diodes from about 0.7–1.6 μm. This range includes important laser applications of fiber communications, fiber sensing and LiDAR.

Dr Robert N Hall from General Electric first demonstrated the semiconductor laser, and filed his patent for the idea in October 1962. Laser diodes at first were homojunction diodes, in which the materials of waveguide core layer and cladding layer were identical. However, laser operation in a homojunction semiconductor is difficult to realize due to the enormous current required to achieve optical gain, especially for room-temperature lasing. In the defined region of a homojunction semiconductor, radiative recombination of electrons and holes occurs. However, electrons and holes can just drift through the junction without recombining. An innovative and more efficient solution to make room-temperature operation possible is the invention of the double heterostructure [3]. In a double heterostructure, semiconductors with different bandgap energy are combined, as illustrated in figure 3.4. A layer of semiconductor with low bandgap energy is sandwiched between two layers of semiconductors with high bandgap energy. When the two semiconductors are brought together, potential barriers are formed which can confine electrons and holes. Either junction between different bandgap semiconductors is a heterostructure. The sandwiched structure is a double heterostructure that further improves the confinement. In the case of a double heterostructure formed by different types of semiconductors in equilibrium, the Fermi level is constant across the junction and causes the band profile to bend. When the double

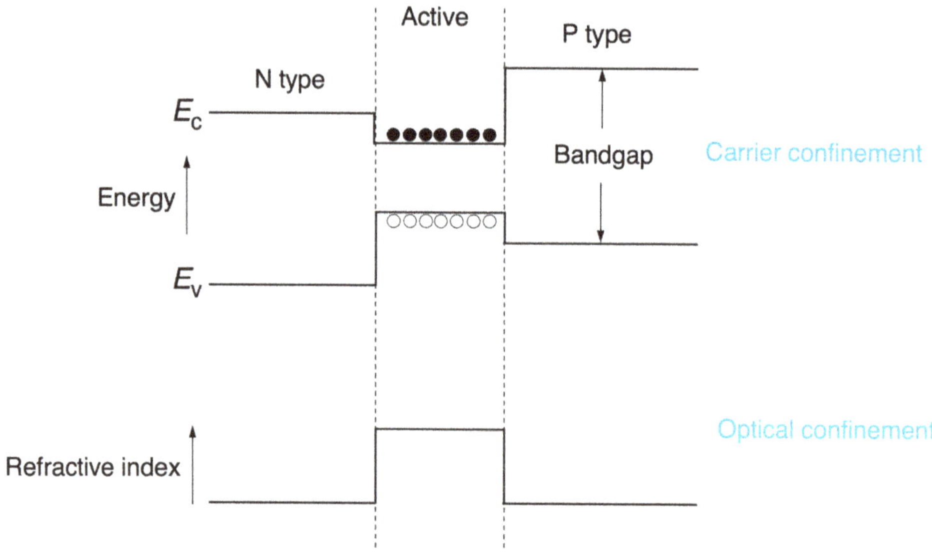

Figure 3.4. Carrier confinement and optical confinement of double heterostructure.

heterostructure is forward biased, an injection of electrons and holes is made into the device, the depletion region is reduced and the band of the n-type semiconductor shifts upwards. If the voltage is sufficient, the quasi-Fermi level for the n-type semiconductor is at the same energy. Then, the electrons can overcome the potential carrier and flow into the active region, where they are confined by the lower bandgap material. Likewise, holes flow into the active region from the *p*-type semiconductor. So, the population inversion is created. The electrons and holes are confined where they can combine radiatively to emit photons. Considerable progress towards room-temperature continuous wave operation was achieved by the introduction of the double heterostructure. The double heterostructure has two significant features of carrier confinement and optical confinement that are essential to laser operation. The active region where free electrons and holes simultaneously exist and recombine confined within the thin middle layer has thickness even in the nm regime, resulting in more electron–hole pairs confined within the active region making contributions to the amplification. On the other side, the double heterostructure also yields optical waveguiding due to the bandgap difference in semiconductors. A lower bandgap semiconductor sandwiched in the middle has a larger refractive index, while the cladding semiconductor with higher bandgap has a smaller refractive index. Thus, refractive index difference makes the active region with high refractive index play as a waveguide core to provide optical confinement and better overlap between modal profile and active region.

Population inversion is not enough to generate a laser. In order to make stimulated emission possess a significant position, the light must interact with the electrons in the conduction band. The interaction can be achieved by placing the active region inside a resonant cavity, in which the light is reflected back and forth multiple times before leaving the cavity. During the round-trip process, the light gets accumulated gain. In the condition that gain equals loss, lasing occurs. The illustration of resonant cavity can be found in figure 3.1. The design of cavity structures for modern heterostructure lasers can be much more complicated incorporating more than one set of cladding layers to confine the carriers (separate confinement heterostructure, SCH) or graded-index separate confinement heterostructure (GRINSCH). Active regions are also created by quantum wells due to their higher efficiency brought by the stepwise form of density of states function.

Compared to other lasers, diode lasers are highly efficient with merits of extremely tiny footprint, bright output considering their small size, low threshold current, and low energy consumption. The simple diode laser described above has been heavily modified in recent years to accommodate modern technology, resulting in two configurations of edge emitting diode laser and surface-emitting diode laser.

3.2 CW and pulsed lasers

Semiconductor lasers generally can be divided into two categories: continuous wave (CW) and pulsed, as illustrated in figure 3.5. Laser output is normally continuous for most semiconductor lasers, due to limited capability of energy storage.

CW laser—A CW laser refers to operation at a steady state with duration longer than 0.25 s [4]. It is considered to pump continuously and emit light continuously. The continuous operation leads to heat accumulation. The consideration of heat removal is significant for a CW laser. As a result, CW laser diodes usually are long enough to ensure effective heat removal over their length and are often mounted on

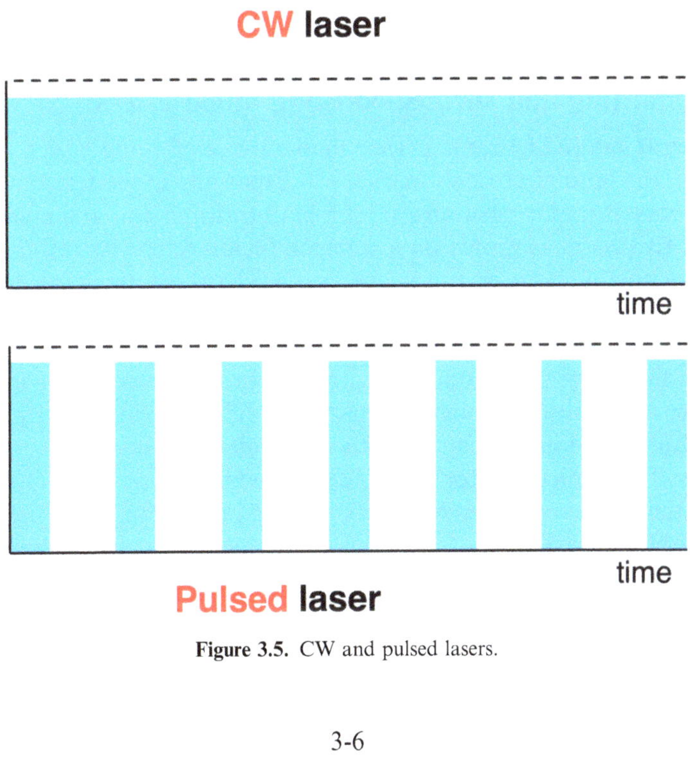

Figure 3.5. CW and pulsed lasers.

a highly thermal-conductive diamond substrate. Output power can be manipulated in accordance to a periodic analog or digital signal. The modulated diode lasers play a key role in optical transmission systems and drive the development of ICT (information and communications technology) infrastructures. Amplitude modulation of a continuous wave (AMCW) or frequency-modulated continuous wave (FMCW) LiDARs rely on the modulation of CW lasers to find the distance of objects. While, time-of-flight (ToF) LiDARs rely on pulsed lasers.

Pulsed laser—With the manner of Q-switched mode locking or gain switching, a pulsed laser emits light not in a continuous mode, but rather in the form of optical pulses. The optical power appears in pulses of some duration at some repetition rate. Due to the low duty cycle below 0.1% of pulsed lasers, heat is not a major issue for a pulsed laser diode. Thus, a pulsed laser design might feature a short resonator mounted directly on the base of a TO-type housing (TO-type housing can be found in the preface photo in this chapter). Q-switched mode locking is an operation regime achieved by placing some type of variable attenuator inside the optical laser resonator. Extremely high peak power or ultrashort pulses can be obtained in Q-switched mode locking lasers, while gain switching is an active method to produce pulses by periodically modulating laser gain via pump power in a short duration. Pulsed emissions are switched on or off by injecting carriers to make carrier density within active regions from below to above the lasing threshold or from above to below the lasing threshold. Gain switching techniques are suitable for the generation of nanosecond or picosecond pulses with relatively low energy. Pulsed laser diodes are designed to be driven with high-current pulses to produce short, high-power optical pusles. The driver circuit will be elaborated in detail in section 2.5. Pulsed laser diodes are widely used in ToF LiDARs which measure distance by counting the round-trip optical time.

3.3 Edge emitting and surface-emitting diode lasers

Edge emitting lasers (EEL) came earlier than surface-emitting lasers, and are well established in the opto-electronics market [5]. Especially in the telecommunications area, EELs possess major positions in long-haul transmission and coherent optical systems. An EEL laser is grown on a semiconductor wafer. Its resonator together with the direction of laser emission are parallel to the wafer. The length of the resonator is typically between hundreds of microns and several millimeters, providing sufficient gain to overcome optical loss. Current flows from the *p*-type semiconductor to the *n*-type semiconductor with electrons and holes being injected in the active region. As mentioned above, the bandgap difference induced by the different semiconductors in a double heterostructure provides confinement for both carriers and photons. The semiconductor with higher bandgap has a lower refractive index than that in the active region, giving index guiding in the transverse direction. In the plain of the active region, the emission is confined by gain guiding, where the refractive index is modified by the carrier density. Light is confined within a waveguide structure formed by the index difference of refraction in a double heterostructure. The double heterostructure simultaneously restricts carriers to a

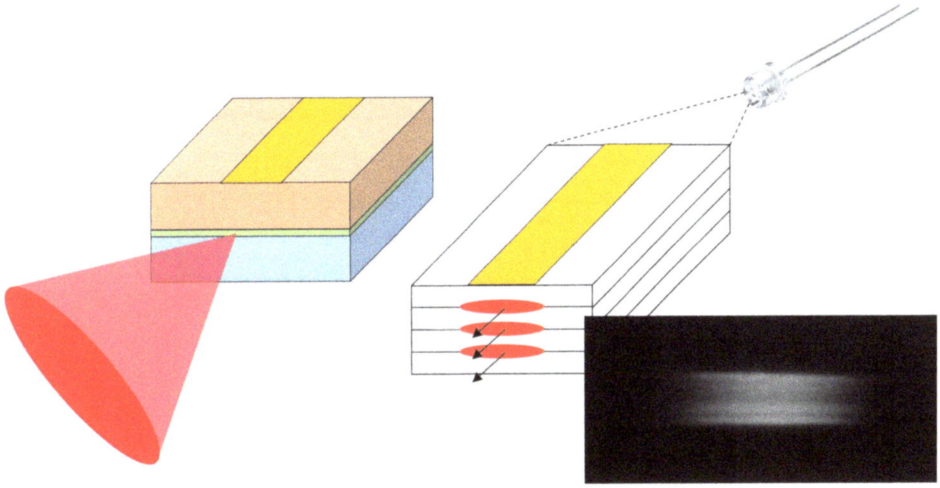

Figure 3.6. Diagram of EEL (left) and stacked EEL (right).

narrow region and serves as a waveguide for optical field, resulting in a high efficiency and lowering threshold. The cleaved end facets provide partial reflection at the semiconductor/air interface by the *Fresnel* reflection for the propagated light within the cavity. The longitudinal waveguide together with two cleaved end facets with reflection form a *Fabry–Pérot* resonant cavity. To further increase the reflection, one end is coated by highly reflected coating with about 95% reflectivity or more. To obtain higher power, the other end stays partially reflective. With longer gain length compared to surface-emitting diode lasers, edge emitting diode lasers are capable of producing laser with higher power. Mass-production laser diodes like SPL PL 90_3 from *Osram* usually increase the output power by stacking multiple active regions (illustrated in figure 3.6). As shown in figure 3.6, the far-field pattern of SPL PL_3 has multiple bright streaks without the collimation of lenses. The implementation of an EEL requires extra components like back facet detectors to stabilize the output power and wavelength in mission-critical applications like coherent optical communications. Due to the diffraction induced by the flat geometry of the active region, the output beam is elliptical rather than circular-like. Therefore, an extra lens or group of lenses is required for beam shaping. With the merits of high peak power and narrow linewidth, EELs are widely used as laser source in pulsed ToF LiDARs and FMCW LiDARs.

Surface-emitting diode laser is a kind of laser diode that emits laser beam in a direction perpendicular from the top surface, which differs from EEL. Vertical-cavity surface-emitting laser (VCSEL) is a widely known surface-emitting diode laser, being used in consumer electronics, telecommunications, and sensors. The resonant cavity of a VCSEL consists of two distributed Bragg reflector (DBR) mirrors parallel to the wafer surface and an active region consisting of multiple quantum wells between the two mirrors, as illustrated in figure 3.7. The thickness of an active region is only a few micrometers, while the thickness of a DBR mirror is

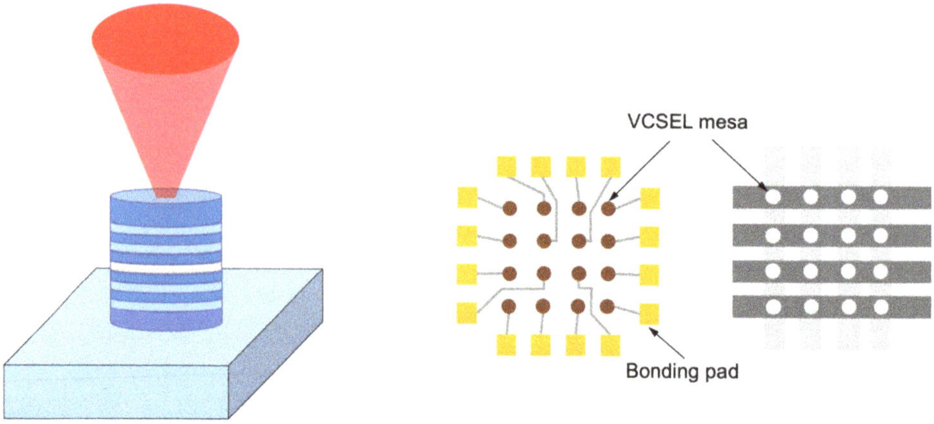

Figure 3.7. Diagram of VCSEL (left) and addressable VCSEL arrays (right).

much larger than that of an active region [6]. The DBR mirror consists of layers with alternating refractive indices to provide high reflection for light within the cavity. To achieve extremely high reflectivity within a wavelength range, the thickness of each layer approximately equals a quarter of the laser wavelength in the material [7]. The output surface is with a slightly low reflectivity by the delicate design of DBR mirror. Due to the geometric design of the surface-emitting, the volume of the active region in the vertical layer is smaller compared to the counterpart of EEL, reducing the threshold current and the power consumption. One of the merits of VCSELs is the high beam quality, compared to EELs. Due to its cylindrical cavity perpendicular to the wafer, the beam profile of a VCSEL is approximately circular with a smaller divergence angle, enabling higher coupling efficiency with optical fibers and easier collimation than an EEL. A significant practical advantage of VCSELs is that they can be tested and characterized *on-wafer* before being cleaved into individual devices. This makes it possible to identify quality issues and reduce the fabrication cost. Furthermore, the surface-emitting approach allows VCSELs to work not only in one-dimensional, but also in two-dimensional arrays. Output power can be significantly increased, by collimating output beams of arrayed VCSELs. In an addressable VCSELs array, output beam can be manipulated individually, by column or by row to realize displays, optical interconnects, and solid-state LiDARs.

3.4 Driver circuit of laser diode

ToF LiDARs measure distance in an active manner by emitting short laser pulses to an optically visible target and measuring the transit time between the emission and the arrival of reflected pulse at the receiver. Laser pulses are generated by semiconductor laser diodes which are usually EELs operating at a wavelength of 905 nm. Semiconductor laser diodes have similar electrical properties to diodes. Laser diodes are low-voltage current-driven devices where a large amount of electrons are required to be injected into diodes for the conversion from electrons to photons. Pulsed laser diodes require injection of abundant electrons in an extremely short time. Under forward voltage of 5–12 V, a high current of up to 40 A is required to deliver the power required. The peak power, rise time and pulse width of emitted pulses are key parameters to determine the maximum ranging capability and ranging accuracy of a LiDAR, which depends on the current supplied to the laser diode by the driver. To obtain long-distance measurement and high SNR, the transient peak power and current need to be typically more than 10 W and 10 A, respectively. Pulsed laser diodes with peak power up to several hundreds of Watts and pulse width down to 1 ns are now off-the-shelf devices, enabling detection range up to 200 m.

There are two primary candidates of driver circuit topologies that provide transient high power: the leading edge controlled resonant laser driver and the current limited dual edge controlled driver [8, 9]. Since the resonant driver is the most common for high-speed applications, the following discussion will be around it. As illustrated in figure 3.8(a), a transmitter driver consists of an energy storage to reserve energy, a switch to control the charge and release of the energy storage and a power supply to provide stable electric power. An energy storage unit is usually a capacitor in the equivalent circuit in figure 3.8(b). A switch unit can be an avalanche transistor, a metal-oxide-semiconductor field-effect transistor (MOSFET) or a GaN FET. To obtain short laser pulses, the gate of the switch must be charged very quickly, i.e., the specific gate capacitance must be charged within the range of a few nanoseconds. Among the above switches, MOSFET is widely used due to its merits of low-voltage control and fast switching. Compared to MOSFETs, GaN FETs have lower gate capacitance, which enables faster switching and therefore shorter pulse width [10]. Moreover, due to the material properties, GaN FETs can stand higher voltage and current. But, the cost of a GaN FET is much higher than a MOSFET, and the complementary metal-oxide-semiconductor (CMOS)-compatible GaN fabrication may not be ready for mass production at present.

This circuit operates in two phases. During the first phase, the switch is turned off, the capacitor is being charged by the power supply through R_1, until its voltage approaches V. The laser diode is disconnected during this phase. In the second

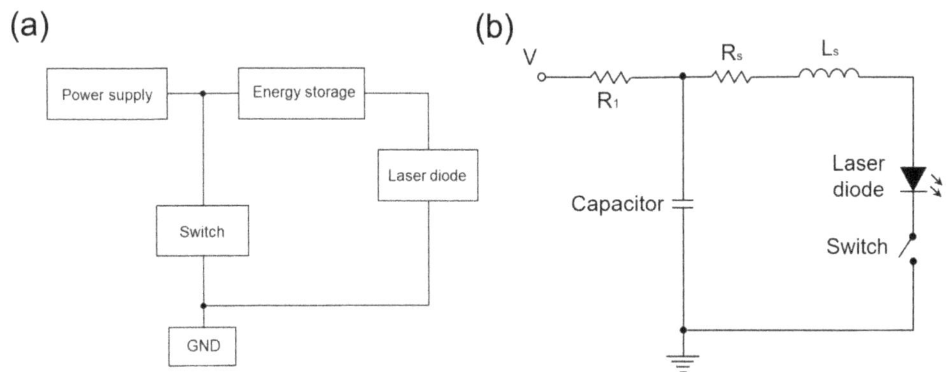

Figure 3.8. Schematics of (a) transmitter driver structure and (b) the equivalent circuit with the key parasitic elements.

phase, the switch is turned on, the charged capacitor begins to discharge rapidly generating a short and high current pulse through the laser diode. So, the laser diode emits a short laser pulse beam with high peak power. Finally, the switch is turned off causing the laser diode to be disconnected and the current pulse to end, then a new charge/discharge cycle begins.

In figure 3.9, assuming Q_1 is an ideal switch and D_L an ideal diode with a fixed forward voltage drop V_{DLF}. In the charging stage, Q_1 starts in the off-state, $i_{DL}= 0$ at this time. The capacitor voltage V_{C1} is being charged to be equal to V_{IN} through R_1. In the discharging stage, when $V_{command}$ triggers the gate driver, the switch Q_1 is fully turned on to discharge the capacitor C_1, letting stored charges flow through the laser D_L and inductor L_1. C_1 and L_1 form a resonant network, hence i_{DL} and V_{C1} oscillate sinusoidally. The effective initial capacitor voltage is $V_{IN} - V_{DLF}$, due to the forward drop of the laser diode. Then, after discharging, i_{DL} returns to zero. At this point, D_L prevents the current from reversing and C_1 recharges via R_1. The charging time constant τ_{chrg} and the resonant period t_{res} are

$$\tau_{chrg} = R_1C_1 \tag{3.1}$$

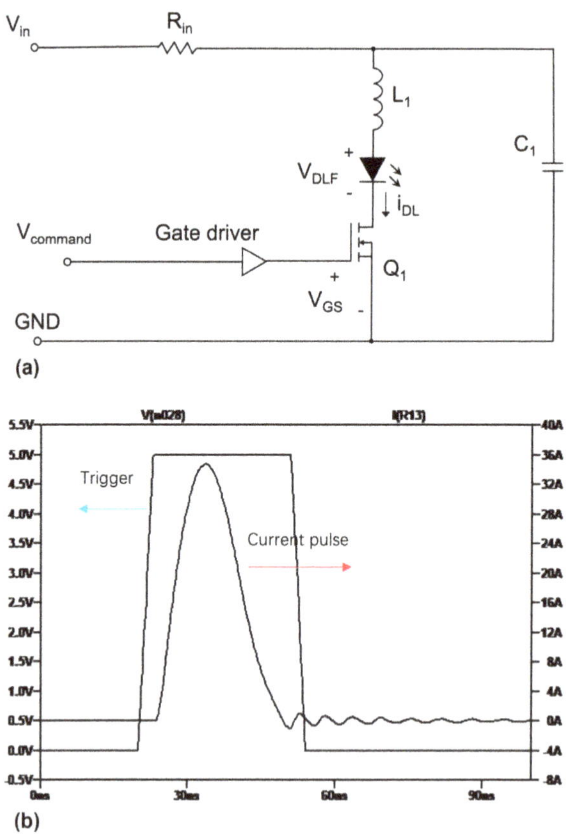

Figure 3.9. Schematic circuit topology of capacitive discharge resonant driver and simulation result of generated current pulse.

$$t_{res} = 2\pi\sqrt{(L_1 C_1)} \tag{3.2}$$

The resonant characteristic impedance R_0, the full width half maximum (FWHM) pulse width t_w, and the peak current of laser diode are

$$R_0 = \sqrt{\frac{L_1}{C_1}} \tag{3.3}$$

$$t_w \approx \frac{t_{res}}{3} \tag{3.4}$$

$$I_{DLK} = \frac{V_{IN} - V_{DLF}}{R_0} \tag{3.5}$$

Therefore, stray inductance has a large effect on laser pulses. The reduction of stray inductance in the transmitter will be elaborated in the next part. The advantages of resonant driver topology include:

- Stable pulse shape;
- Peak power can be set by V_{IN};
- Simple control and drive;
- Pulse width can be shorter than that of $V_{command}$.

A layout of LiDAR transmitter using a MOSFET as switch and the waveforms of the emitted pulse are illustrated in figures 3.10 and 3.11. An SPL PL 90_3 laser diode from *Osram Opto Semiconductors* is adopted due to its outstanding performance and low cost [11]. It operates at a wavelength of 905 nm. The pulse width ranges from 1 to 100 ns with maximum duty cycle of 0.1%. The typical peak output power reaches 75 W owing to the Nanostack laser technology including three epitaxially stacked emitters. A MOSFET of BSC159N10LSFG from *Infineon* is adopted as the switch. UCC27511 from *Texas Instruments* is used to drive the MOSFET [12].

As discussed above, the stray inductance in the current loop including parasitic inductance distributed in the printed circuit board (PCB) limits the rise time of the current injected. The laser diode contributes most of the inductance. The majority of the inductance comes from the laser package, including wire bonds. One way to decrease the laser inductance is to make the laser diode pins very short by soldering them directly onto the PCB. To decreases parasitic inductance, the laser diode needs to be integrated with the driver on the same PCB, and PCB traces also need to be considered. It is important to be noted that thinner PCB with wider traces would contribute less inductance. Multilayer ceramic capacitor (MLCC) capacitors and resistors with smaller package and flipped mounting used in the main current loop will reduce the stray inductance. Parallel alignment of resistors can further reduce the inductance. A Schottky barrier diode of BAT46WJ from *Nexperia* is used as a clamping diode for the protection of the MOSFET and laser diode, against the voltage transients caused by the cutoff of the high loop current. Voltage from RF_1 can be used to monitor the operation of the transmitter (figure 3.11).

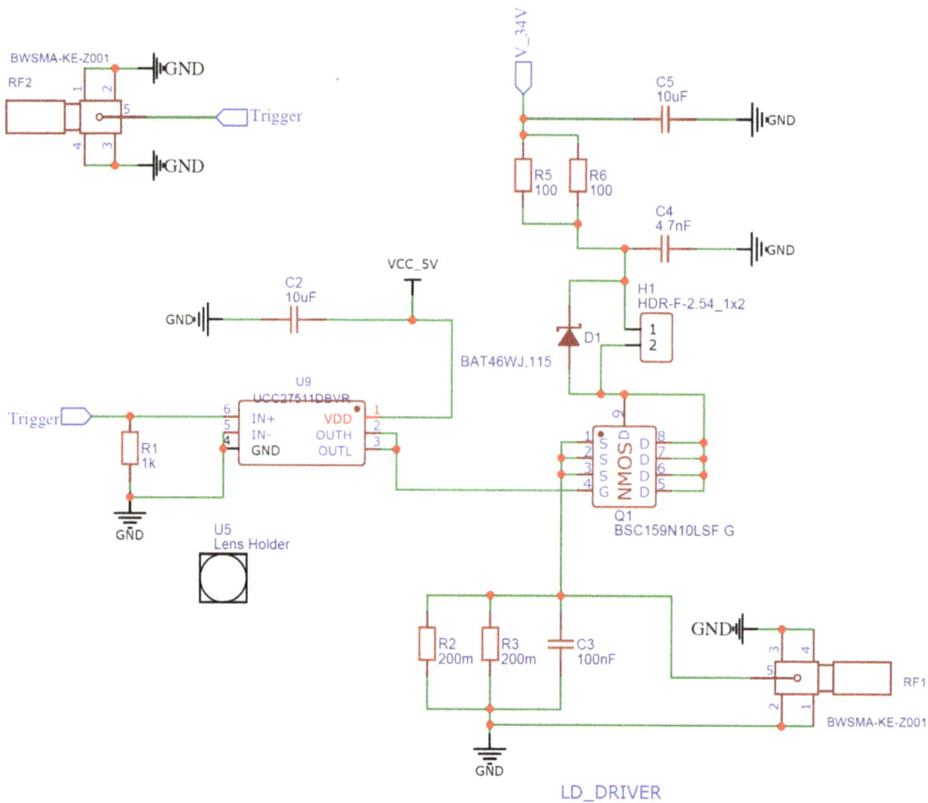

Figure 3.10. Layout of transmitter.

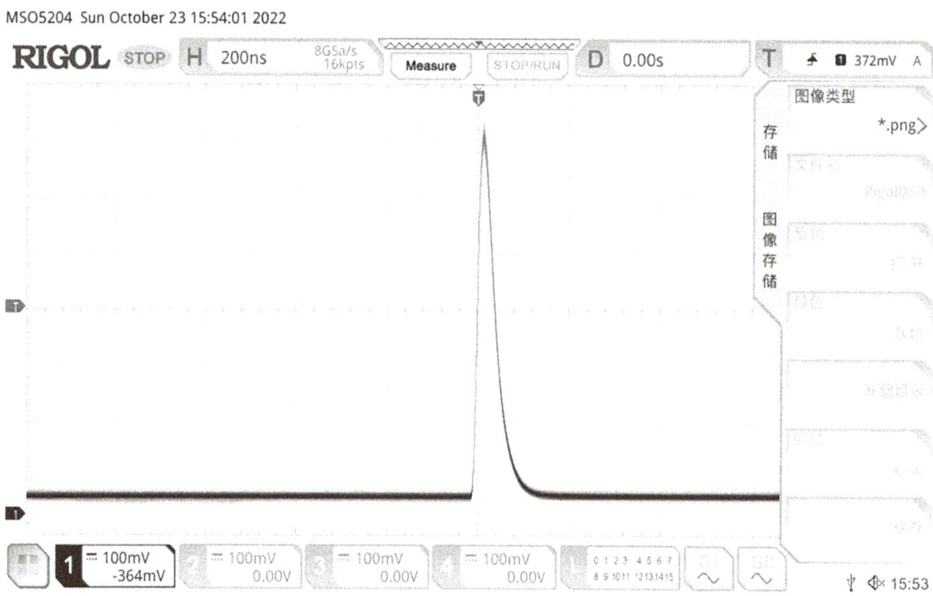

Figure 3.11. Waveform of laser pulse from the above transmitter.

Although the above-mentioned approaches have been discussed to minimize the stray inductance induced by individual components and soldering, an integrated driver chip packaged with the laser diode can be an easy but not flexible approach to reduce inductance. The semiconductor pulsed laser diode of SPL LL90_3 from *Osram Opto Semiconductors* is a hybrid packaged laser module. The hybrid laser module contains two capacitors and a MOSFET acting as a driver stage to provide short and high current pulses for the laser diode. The module is triggered by a transistor–transistor logic-level voltage signal, and is power supplied by a DC voltage.

References

[1] Coldren L A, Corzine S W and Mašanović M L 2012 *Diode Lasers and Photonic Integrated Circuits* 2nd edn (New York: Wiley) ch 2–4

[2] https://en.wikipedia.org/wiki/Optical_cavity (accessed in March 2021)

[3] https://en.wikipedia.org/wiki/Laser_diode#Double_heterostructure_lasers (accessed in March 2021)

[4] https://rp-photonics.com/continuous_wave_operation.html (accessed in June 2021)

[5] https://dentonvacuum.com/comparing-edge-emitting-lasers-and-vcsels/ (accessed in May 2021)

[6] https://rp-photonics.com/vertical_cavity_surface_emitting_lasers.html (accessed in May 2021)

[7] https://en.wikipedia.org/wiki/Vertical-cavity_surface-emitting_laser (accessed in June 2021)

[8] Abramov E, Evzelman M, Kirshenboim O, Urkin T and Peretz M M 2020 Low voltage sub-nanosecond pulsed current driver IC for high-resolution LiDAR applications *IEEE J. Emerg. Sel. Top. Power Electron* **8** 3001–13

[9] Svilainis L, Chaziachmetovas A, Eidukynas V, Aleksandrovas A and Varantinskas M 2021 Compact laser driver for ultrasonic arbitrary position and width pulse sequences generation *IEEE Trans. Instrum. Meas.* **70** 7006815

[10] Glaser J S *Application Note* **AN027** eGaN FETs for lidar—getting the most out of the EPC9126 laser driver

[11] Operating the Pulsed Laser Diode SPL LL90_3 *Application Note* **AN105** https://dammedia.osram.info/media/resource/hires/osram-dam-2496583/Operating%20the%20pulsed%20laser%20diode%20SPL%20LL90_3.pdf (accessed in May 2021)

[12] Renesas Electronics *Application Note* **AN1108** Applying Power MOSFET Drivers https://intersil.com/content/dam/Intersil/documents/an11/an1108.pdf (accessed in March 2021)

IOP Publishing

ToF LiDAR for Autonomous Driving

Wei Wei

Chapter 4

Photodiode and receiver

This Avalanche photodiode image has been obtained by the author from the Wikimedia website where it was made available by Radovan Blažek under a CC BY-SA 3.0 licence. It is included within this article on that basis. It is attributed to Radovan Blažek.

After illuminating laser pulses on targets, the reflected pulses need to be collected and processed by LiDARs to obtain distance information. The returned optical signal is converted to electronic signal by a photodiode, and then its amplitude and arrival time are amplified and discriminated by receiver circuits. The detection and ranging capabilities of a LiDAR significantly rely on the performance of the receiver. In this chapter, the theory of photoelectric conversion in semiconductors is introduced. Then, we get deep into key devices of PIN and avalanche photodiode (APD). Finally, receiver circuits including transimpedance amplifier (TIA) preamplifier, timing discriminator and time-to-digital converter (TDC) is elaborated.

4.1 Fundamentals of photodiodes

A photodiode is an opto-electric semiconductor device that absorbs photons and then produces electric current. The photodiode works on the photoelectric effect that was discovered by the German physicist Heinrich Rudolf Hertz in 1887 [1, 2]. The production of electrons when light hits the material is known as the photoelectric effect. The produced electrons are known as photoelectrons. The photoelectric effect can be categorized as the external photoelectric effect and the internal photoelectric effect, depending on whether the produced electrons are emitted from the material after photons being absorbed, as illustrated in figure 4.1. The latter is the operating mechanism of semiconductor photodiodes. In the internal photoelectric effect, the produced photoelectrons cannot be observable outside the semiconductor material, but only electrons are excited from the valence band to the conduction band. One important prerequisite is that the photon energy is larger than the bandgap of the material in the active region. The bandgap in each semiconductor material is different, exhibiting diverse absorbing wavelengths. For example, silicon-based photodiodes absorb photons of wavelengths up to roughly 1.1 μm, while indium gallium arsenide with lower bandgap allows absorption at longer wavelengths up to approximately 1.7 μm [3]. Some materials like mercury cadmium telluride have even smaller bandgap, enabling detection in mid-infrared region for applications of infrared cameras.

A photodiode is a type of diode that contains a PN junction, its symbol is shown in figure 4.2. As presented in figure 4.3, a PN photodiode has two layers of $P+$ and $N-$, between which is a depletion region. On the top of the P layer is an anti-reflection coating to increase the collecting efficiency of a photodiode. The P layer has an abundance of holes, and the N layer has an abundance of electrons. The depletion region that has few free charge carriers is formed by the diffusion of electrons from N type layer to P type layer and the diffusion of holes from P type layer to N type layer. The P and N type layers are thin, and most of absorption exists in the depletion region, where photodetection occurs. The depletion region formed between the two layers acts as a capacitor, where the boundaries of the $P+$ layer and $N-$ layer act as parallel plates of the capacitor formed inside the junction. Its

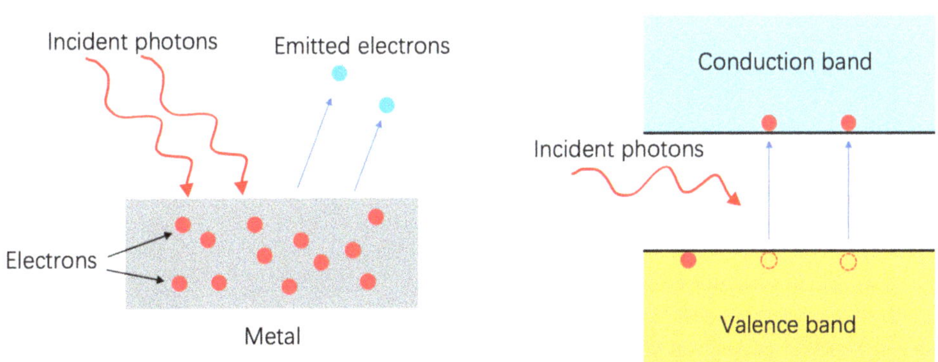

Figure 4.1. External (left) and internal (right) photoelectric effects.

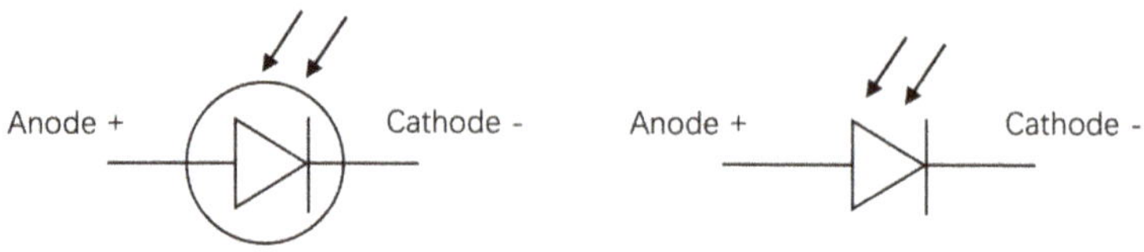

Figure 4.2. Symbols of photodiode.

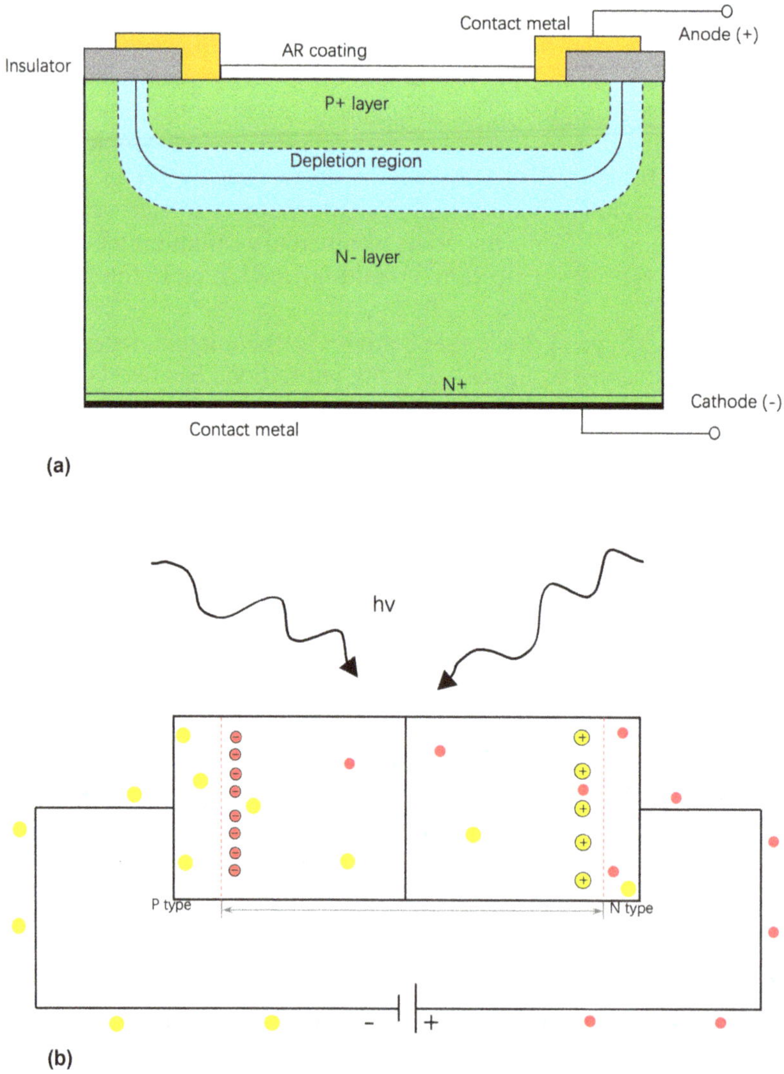

Figure 4.3. Schematic diagrams of photodiode.

capacitance is inversely proportional to the width of the depletion region. The region develops a built-in voltage to create an electric field across the depletion region allowing the current to flow through anode–cathode in one direction. When a photon with sufficient energy strikes the diode, an electron–hole pair is created. If the absorption occurs in the depletion region, carriers are swept from the junction by the built-in electric field of the depletion region. Holes move towards the anode, and electrons move towards the cathode. These moving carriers form photocurrent that flows only in one direction from the anode to the cathode. Different from a laser diode, a photodiode operates in reverse bias condition rather than forward bias. When light falls on it, the amount of current flow is directly proportional to the intensity of light under reverse bias. The total current through the photodiode is the sum of the dark current and the photocurrent. The dark current refers to a very small amount of current in condition of no illumination or light on photodiode that is under reverse bias [4]. To maximize the sensitivity of a photodiode, the dark current must be minimized. However, the width of the depletion region in a PN photodiode is small, leading to low sensitivity to incident light. PINs and APDs adopt wider depletion region or avalanche effect to address this issue, which will be introduced in the following sections.

A photodiode can operate either as a photovoltaic or photoconductive device, as shown in figure 4.4.

Photovoltaic mode: In photovoltaic mode, the photodiode operates under no bias without being connected to any power source. When light illuminates it, electrons inside the photodiode are excited to higher energy state, resulting in the movement of electrons towards the cathode and holes towards the anode. This process creates a potential difference between two terminals, making the photodiode a current source. This mode exploits the photovoltaic effect, which is the basis for solar cells. To get a large amount of illumination, a solar cell usually has a much larger photosensitive area.

Photoconductive mode: In photoconductive mode, the photodiode is connected to a power source and reversely biased, that is, the cathode is driven positive with respect to the anode. When light illuminates the photodiode, pairs of electrons and holes are created and move towards the opposite direction due to biased voltage. The applied reverse bias increases the width of the depletion layer, bringing two beneficial effects. First, a wider depletion region makes the photodiode more sensitive to incident light. Thus, the photoconductive mode is a wise option to produce stronger signal relative to illuminance. Second, a wider depletion region reduces the junction capacitance of the photodiode. As with a basic RC low-pass filter, reducing capacitance increases the cut-off frequency. Thus, the photoconductive mode allows for wider bandwidth and is preferable when you need to increase the detector's capability to respond to faster variations of illuminance. This mode is the operating mode to get the variation of signals in optical communications and LiDAR.

Critical performance-related metrics of photodiodes include quantum efficiency, responsivity, gain, noise-equivalent power, detectivity, dark current, response time and noise spectrum. The ratio of light that is transduced into electrical signal is determined by quantum efficiency and responsivity. Gain describes the amplification

(a)

(b)

Figure 4.4. Solar cells versus photodiodes. (a) This Silicon heterojunction solar cell image has been obtained by the author from the Wikimedia website where it was made available by Radiotrefoil under a CC BY-SA 4.0 licence. It is included within this article on that basis. It is attributed to Radiotrefoil. (b) This NORTEL AA1419001 Optical interface tear down image has been obtained by the author from the Wikimedia website where it was made available by Mister rf under a CC BY-SA 4.0 licence. It is included within this article on that basis. It is attributed to Mister rf.

of incident light by inherent multiplication of a photodiode. The speed of a photodiode is characterized by response time. The noise in transduced electrical signal is described by noise-equivalent power, detectivity, dark current, and noise spectrum, among which noise spectrum characterizes frequency-related noise [5].

Quantum efficiency is defined as the percentage of incident photons that can be converted into electrons by a photodiode. It is classified by two categories of internal quantum efficiency and external quantum efficiency. The internal quantum efficiency is referred to as the ratio of the number of charge carriers generated to the number of photons absorbed within the active layer of the device. It is very high and can be close to 100% in a superior quality material with low dislocation density and defects, while the external quantum efficiency considers only the photogenerated carriers than can be collected or measured as a result of light absorption. It is lower than the internal quantum efficiency in any photodiode due to the factor that not all photogenerated pairs manage to come out of the device as electrical current. As mentioned above, external quantum efficiency can be defined by the ratio of carriers collected to all incident photons

$$\eta = \frac{i_{ph}}{q} \bigg/ \frac{P_{opt}}{hv} = \frac{i_{ph}}{P_{opt}} \frac{hv}{q} \tag{4.1}$$

where i_{ph} is the photocurrent generated in response to incident light, q is the electron charge, P_{opt} is the incident optical power in Watts. hv denotes the energy of the photon. Here, i_{ph}/P_{opt}, the ratio of photocurrent to input optical power, is a fundamental figure of merit for any photodiode, and is defined as the intrinsic responsivity or **spectral responsivity** if plotted with respect to wavelength, with unit of A/W (figure 4.5).

Noise-equivalent power (NEP) is the common metric that quantifies a photodiode's sensitivity or the power generated by a noise source. It is the amount of light

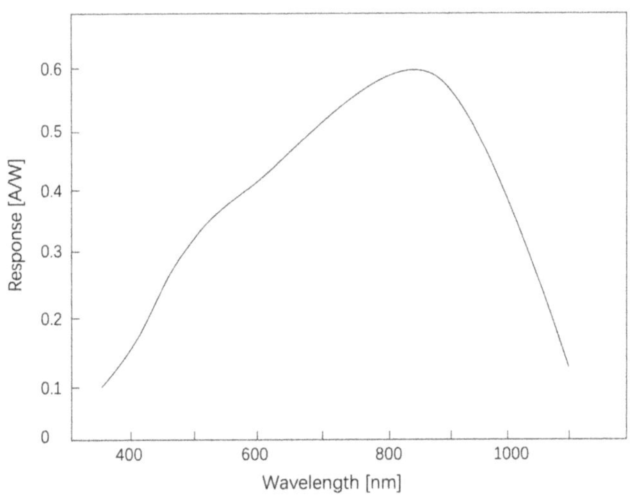

Figure 4.5. Response of a silicon photodiode as a function of incident wavelength.

required to collect a signal equivalent in power to that of the noise in a photodiode device, i.e., the input power that results in a signal-to-noise ratio (SNR) of 1 in a 1 Hz output bandwidth. For photodiodes, the NEP expresses the sensitivity of the device and is given in Watts per square root of Hertz ($W/\sqrt{Hz}$). It is the minimum detectable power per square root bandwidth of a given detector. Therefore, it is desirable to have an NEP as low as possible, since a low NEP value corresponds to a lower noise floor and therefore a more sensitive detector. Even at higher input light intensities, a low NEP is beneficial since it will lead to lower noise characteristics in the output signal. The minimum detectable power can be calculated by

$$P_{\min} = \mathrm{NEP}(\lambda) \times \sqrt{\mathrm{BW}} \tag{4.2}$$

where NEP(λ) is the wavelength-dependent NEP and BW is the measurement bandwidth (figure 4.6).

Detectivity is a merit defined as the inverse NEP and detective area. The larger the detectivity of a photodiode, the more it is suitable for detecting weak signals which compete with the detector noise. It can be calculated as

$$D = \frac{\sqrt{A}}{\mathrm{NEP}} \tag{4.3}$$

Gain characterizes the amplification of photogenerated current in a photodiode and is defined as the ratio of total current out of the device to the photogenerated current

$$G = \frac{I_{\mathrm{out}}}{i_{\mathrm{ph}}} \tag{4.4}$$

In PN or PIN photodiodes, the maximum gain is less than 1, since there is no carrier multiplication in devices. In APDs, multiplication can be multiple times or tens of times, which results in gain much larger than 1. In the meantime, the noise in devices also can be amplified by the carrier multiplication.

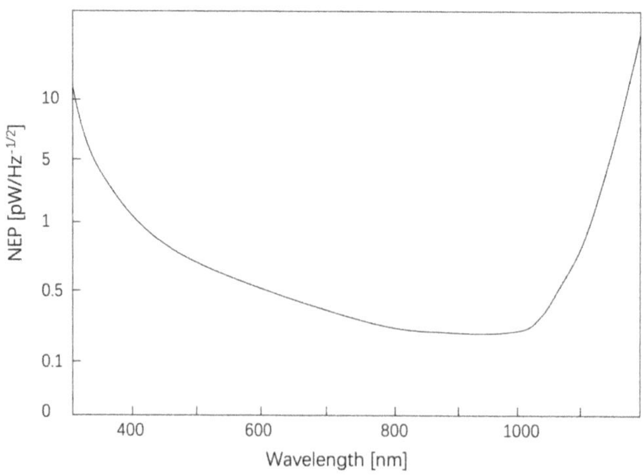

Figure 4.6. Plotted NEP as a function of wavelength.

Dark current is a tiny amount of signal current in a photodiode under reverse bias even in the absence of any incident light. It is unwanted leakage current in a photodiode. Dark current is often caused by thermal excitation of carriers, not necessarily from valence to conduction band, but possibly through defect states related to crystal defects or impurities. The rate of thermal process depends not only on the active area, but also the temperature and bandgap energy of the material, and the operation voltage. Dark current cannot be avoided, but can be suppressed by increasing the level of signal to achieve maximum sensitivity and SNR.

Response time is an important characteristic of a photodiode, expressing the time that it takes for the photodiode output to change in response to changes in the incident light intensity. It is also referred as rise time, which is the time required for a photodiode to go from 10% to 90% of its final response state. The response time is usually dependent either on the size of active area or the RC time constant of circuit. It is also dependent on how long it takes a photogenerated electron to travel from the active area to one of the output terminals.

4.2 PIN

PN structure photodiodes suffer from shallow depletion region, leading to limited quantum efficiency and detection bandwidth [6]. The width of the depletion region in a PN photodiode may be well below the absorption length, as illustrated in figure 4.7. In this condition, only part of the generated photocarriers are generated within the depletion region, while the collection of those generated outside the depletion region may be limited. Consequently, the quantum efficiency is reduced. Even though the carriers generated outside the depletion region eventually diffuse into the depletion region and can thus contribute to the photocurrent. The duration of that diffusion results in a tail in the impulse response function, limiting the detection bandwidth.

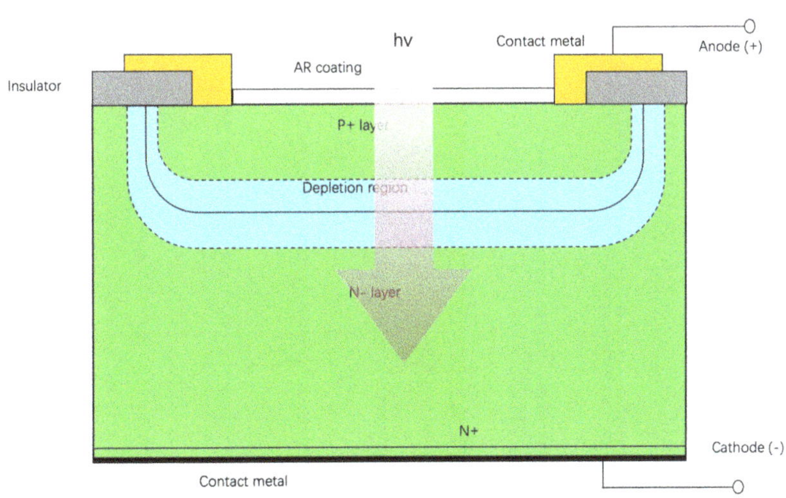

Figure 4.7. Absorption length in PN photodiode.

Hence, PIN structure is proposed to solve the above problems. A PIN photodiode is similar to a PN photodiode with one major difference. Instead of placing *P* and *N* layers together to create a depletion region, an intrinsic layer is placed between the two doped layers to overcome the disadvantage of the small depletion region in PN photodiode, so that it refers to PIN photodiode, as illustrated in figure 4.8. It was invented by Jun-Ichi Nishizawa and his colleagues in 1950. The *P* type and *N* type regions are typically heavily doped, while the wide *I* type region between the *P* type and *N* type regions stays undoped intrinsic [7]. Similar to the PN photodiode, the operation of PIN photodiode is also under reverse bias. The intrinsic *I* type region is flooded with charge carriers from *P* type and *N* type regions. The photodiode will conduct current once the flooded electrons and holes reach an equilibrium point, where the number of electrons is equal to the number of holes in the intrinsic region. The issues of quantum efficiency and bandwidth can be mitigated in a PIN photodiode. Most carriers are generated in the intrinsic region, since that is much thicker than the depletion region in a PN structure. Moreover, the thick intrinsic region reduces the capacitance, which allows for a higher detection bandwidth. Due to the large size of intrinsic region, the sensitivity of a PIN photodiode is more than a regular PN photodiode. Its response is also faster than a PN photodiode. The fastest PIN photodiodes have bandwidths well above 100 GHz. The diameter of active areas are typically only a few microns for optical communications, and can be above several hundred microns for LiDARs. Even for the photovoltaic mode in solar cells, the PIN junction also outperforms regular PN junction in long-wavelength response. In the case of long-wavelength irradiation, photons penetrate deep into the cell. But only those electron–hole pairs generated in and near the depletion region contribute to current generation. The wider depletion width in the thick intrinsic region enables electron–hole pair generation deep within the device, increasing the quantum efficiency of cells.

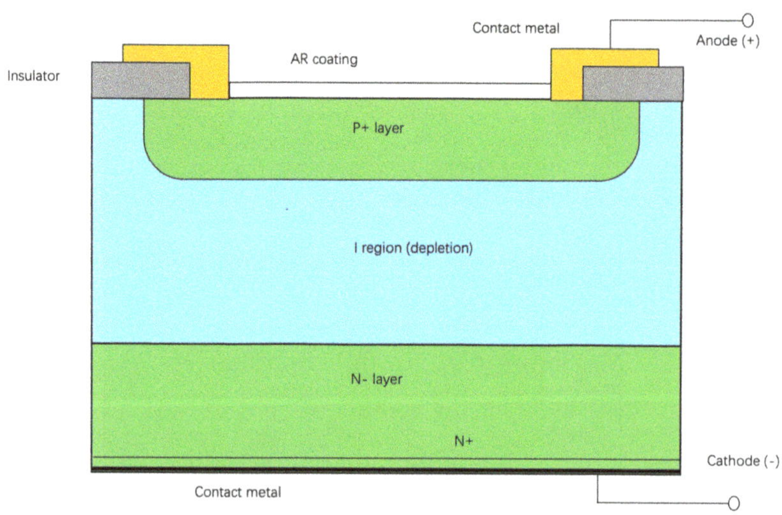

Figure 4.8. Cross-section of PIN photodiode.

4.3 APD and SPAD

Similar to a PIN photodiode, an APD consists of three regions: *P* type, *N* type and intrinsic regions, as illustrated in figure 4.9(a) and (b). The difference is that reverse bias applied is very large to cause impact ionization [8]. The impact ionization results in the avalanche multiplication of photogenerated electrons, providing internal gain, as illustrated in figure 4.9(c). Therefore, the sensitivity of APD is much higher than that of a PIN photodiode. Typical semiconductor APD operates with a relatively high reverse voltage in the tens or even hundreds of volts, sometimes

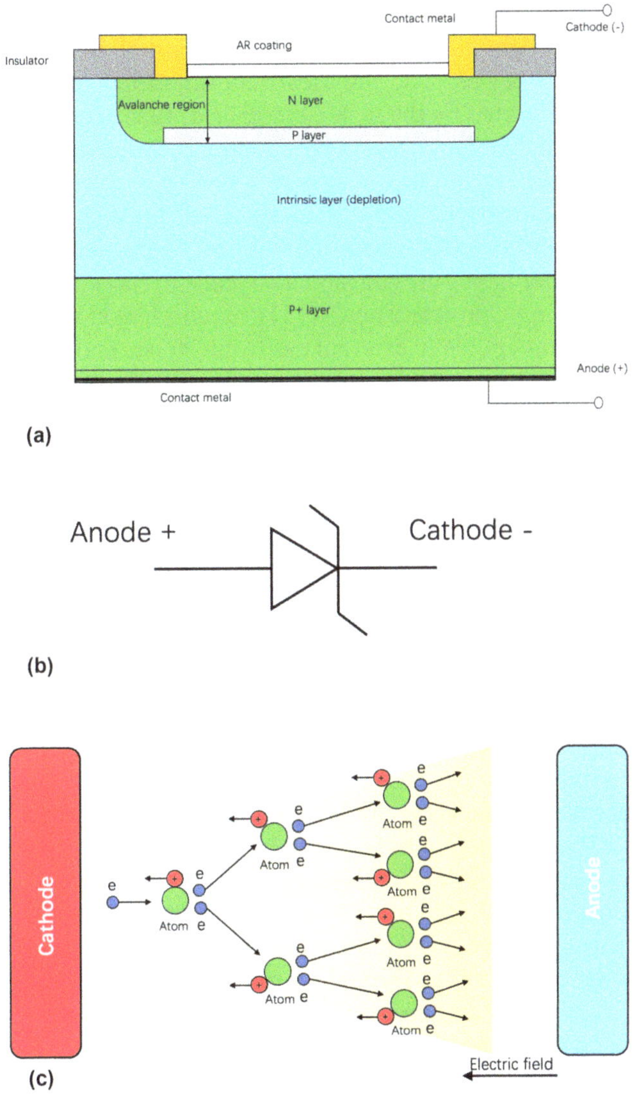

Figure 4.9. Cross-section, symbol and avalanche process of APD.

Table 4.1. Comparison of photodiodes.

Device type	Reverse bias (V)	Gain	Responsivity (A W^{-1})	Response time (ns)	Dark current (nA)
PN	<100	$\leqslant 1$	0.3–0.6	150–2500	0.002–0.2
PIN	<100	$\leqslant 1$	0.14–0.7	10–10^6	0.001–10
APD	100–200 or higher	50–100 or higher	25–50	>0.003	0.05–30

just below breakdown. For a silicon APD, the reverse bias voltage is usually between 100 and 200 V. When light falls on the avalanche region of an APD, the generation of electron–hole pairs begins. The photocarriers migrating toward the avalanche region are highly accelerated by the high electric field, thus acquiring enough energy to ionize lattice atoms [9]. This creates the secondary carriers and generates more carriers via the avalanche multiplication process. Owing to the internal amplification mechanism, high photocurrent and enhanced SNR are achieved. Compared to receivers with PIN photodiodes, those that utilize APDs achieve 5–10 dB better sensitivity. On the other hand, the collision also leads to more heat generation. As a result of their operating characteristics, APDs require less electronic signal amplification and are less susceptible to external electronic noise, making them useful with extremely sensitive detectors. APDs are presently the photodetector of choice in high-speed, short-pulse, and ultra-low detection applications due to the high gain-bandwidth product compared to regular PIN-based detectors. Comparisons of PN, PIN, and APD are presented in table 4.1.

Except for the above-mentioned linear mode that is similar to a PN or a PIN photodiode, an APD can also operate in the so-called *Geiger* mode even for single-photon detection with well-designed electronics. The *Geiger* mode is the mode in which the diode operates slightly above the breakdown threshold voltage, where a single electron–hole pair can trigger a strong avalanche. Thus, the key difference between the linear and *Geiger* mode is the operating voltage, as shown in figure 4.10. The former is below the breakdown voltage, while the latter operates above the breakdown voltage. In the case of the *Geiger* mode, an APD is very easily saturated even by one or several photons. Therefore, an external electronic quenching circuit is required to reduce the voltage at the diode below the threshold voltage for a short time and then increase it to the desired level, which is referred to as quenching circuit. This is so that the avalanche stops and the detector gets ready for the next detection of photons after some recovery time like 100 ns. This insensitive period is referred to as dead time, and this dead time constitutes a substantial limitation on the bandwidth of APDs. It limits the count rate to the order around 10 MHz, whereas the operating bandwidth is from tens to hundreds of GHz for the linear mode.

Photon-counting APDs operating in the *Geiger* mode are also referred as SPADs (single-photon avalanche diodes). SPADs can achieve a gain of 10^6 that is significantly higher than APD, allowing detection of extremely weak light at long

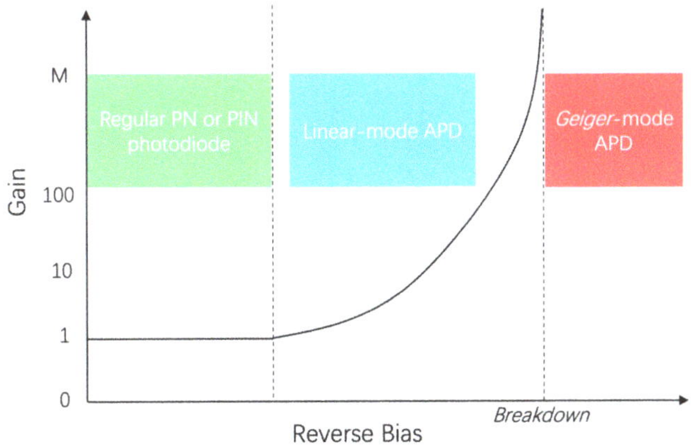

Figure 4.10. Gain of photodiodes versus reverse bias.

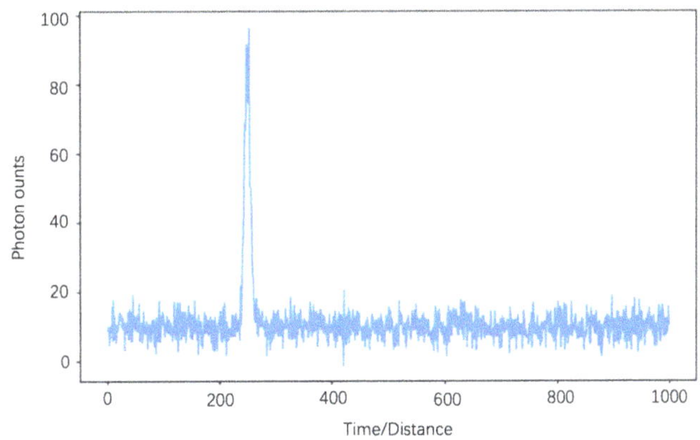

Figure 4.11. TCSPC hologram.

distance. SPADs are also employed in pulsed LiDAR receivers to improve their sensitivity and increase their operating distance. However, SPADs are susceptible to false detections due to either the thermal noise of the detector itself or photons from the ambient light that happens to be at the detectable wavelength window [10]. Therefore, SPAD receivers are often employed in a statistical manner where the arrival times of multiple repetitive pulses recorded by many SPADs are in parallel or array, and then accumulated in a histogram. The recording time window is comparable to the duration of the emitted pulse width to get a higher chance of being part of the expected signal [11]. This technique is referred to as time-correlated single-photon counting (TCSPC), which is also a ranging manner for pulsed LiDAR. A TCSPC hologram example is shown in figure 4.11. In this book, we mainly focus on the time-of-flight (ToF) ranging technique. Arrayed SPADs-based LiDAR is capable of imaging depth information similar to CMOS image sensor which captures color images. Spatial depth information can be obtained simultaneously from arrayed pixels, resulting in solid-state LiDAR with no mechanical scanners or electromagnetic scanners. Solid-state LiDAR enables very compact size

Figure 4.12. LiDAR on iPhone 12 Pro.

to find applications in consumer electronics. The iPhone 12 pro and pro max released in October 2020 have a solid-state LiDAR mounted on the back close to cameras to extend their applications to the 3D world (figure 4.12).

4.4 TIA

Although the APD in a receiver that operates in linear mode has a large gain, the power of returned pulse is still extremely weak if the target is a long distance from the receiver. A weak signal cannot be measured with high accuracy by measurement instruments or electronic circuits. Therefore, an amplifier is required to amplify the signal to higher level for the following processing. However, the output signal from the APD or any other photodiode is current signal, which is not suitable to be processed by electronic circuit. The current signal should be converted to an amplified voltage signal once output from the diode, thereafter cascaded voltage amplification stages may be added to achieve the required signal voltage-magnitude and total circuit bandwidth. There are two approaches to realize current-to-voltage conversion: passive and the active ones. As we know according to Ohm's Law, when current flows through a resistor, it creates a voltage drop across the resistor which is proportional to the value of current and the value of resistor itself. This is the most basic current-to-voltage converter and also a passive current-to-voltage converter. However, a bare resistor as a current-to-voltage converter has low frequency bandwidth if there is significant input capacitance, which there often is. Depending on this situation, we may be concerned with maximum signal transfer and minimizing interstage loading. Both of these answers boil down to the impedance mismatch, which kills signal integrity. Adding an op-amp allows us to

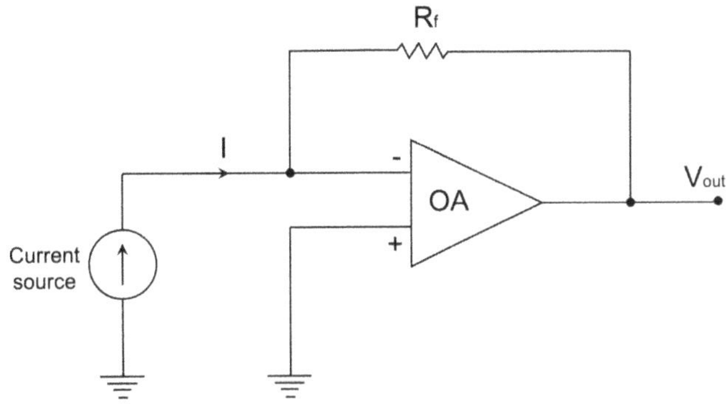

Figure 4.13. Topology of TIA circuit.

control both the input impedance and output impedance. The approach of current-to-voltage conversion built on an op-amp is referred as a transimpedance amplifier (TIA). As an op-amp is an active component, this approach is an active current-to-voltage conversion. In a TIA, we can configure the bandwidth and the gain response of the circuits as per the requirements. It is widely used for the conversion and amplification of signals output from various sensors.

The transimpedance in a TIA is rather than an ordinary impedance, because the current at one place in the circuit becomes a voltage elsewhere in the circuit, rather than having the voltage affect the original current as happens in an ordinary impedance [12]. The circuit diagram of a basic op-amp TIA is shown in figure 4.13. It is a simple inverting amplifier with negative feedback and the op-amp's non-inverting input grounded. The feedback resistor is connected between inverting input and output [13]. The output of photodiode can be considered as current source. We know that the input current of an op-amp at the inverting node will be zero due to its high input impedance. Hence, the input current flows entirely through the feedback resistor, and the op-amp adjusts its voltage output to keep its inputs at equal voltages. The output voltage can be calculated by $V_{out} = -I \times R_f$. The transimpedance of amplifier is simply R_f. We might say it is $-R_f$, but generally the negative sign is not included, because the sign simply indicates the direction of the current and does not affect the behavior of the circuit. When designing a TIA, only the resistance value of the feedback resistor needs to be considered following the constraints on both sides:

1. If R_f is too large, then the current signal will saturate the op-amp's output at either its positive or negative supply rail limits, causing clipping of the signal.
2. If R_f is too small, then the current signal will turn into too small a voltage signal to be useful.

Our goal is to measure the input current, hence creating a larger change in V_{out} to measure by using an R_f that is sufficiently large enough will make it easier to detect small changes in the input current. Therefore, we typically want to choose the largest possible resistance that just barely lets us cover the full range of input

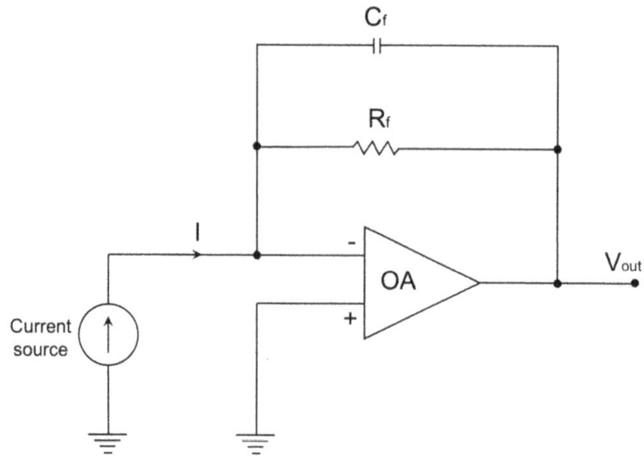

Figure 4.14. Topology of TIA circuit with phase compensation.

current we want to measure. For example, it the op-amp is powered by ±5 V and the current to be measured up to ±100 μA, then we should choose a resistance about $R_f = \frac{5V}{100\mu A} = 50k\Omega$.

This formula will hold true in an ideal circuit. The DC input impedance of the transimpedance amplifier is approximately zero. However, when considering higher frequency effects, it would be wrong to assume the input impedance remains zero at higher frequencies because it actually rises drastically. When a large impedance is combined with even a tiny amount of capacitance, the result is a large time constant $\tau = RC$, significantly reducing the overall bandwidth of the amplifier circuit. In a real circuit, the op-amp includes small value of input capacitance and stray capacitance across its input pins which could cause output drift and ringing oscillation, making the entire circuit unstable. To overcome this problem, instead of a single passive component, another passive component is required to provide phase compensation for the proper working of the transimpedance circuit. The additional component is a capacitor that is connected in parallel to the resistor between the amplifiers inverting input and the output, as illustrated in figure 4.14. This capacitor is also referred as compensation capacitor. The op-amp here is again connected in negative feedback condition through the resistor and the capacitor as the feedback. The current I applied to the inverting pin of the TIA will be converted into equivalent voltage on the output side as V_{out}. The output voltage is not only dependent on the feedback resistor, but it also has a relationship with the value of the feedback capacitor.

The frequency response of TIA is inversely proportional to the gain set by the feedback resistor. While in real situations, the photodiode that is amplified by the TIA usually has an additional capacitance. Its equivalent circuit is shown in figure 4.15. The ideal current source I_P represents the generated photocurrent in response to incident light. The direction of the photocurrent corresponds to current that flows from the diode's cathode to the diode's anode. The photodiode is under reverse bias, and the produced current flows in the direction opposite to that of regular forward-biased diodes. The parallel capacitor C_J represents the photodiode's

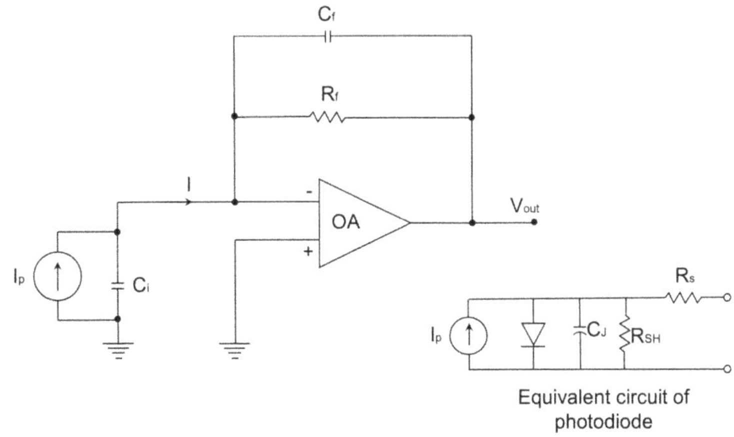

Figure 4.15. Equivalent circuit of photodiode.

junction capacitance, i.e., the capacitance associated with the width of the depletion region in the PN junction, which is mentioned in the above sections. Junction capacitance is a significant parameter since it strongly influences the photodiode's frequency response. Lower junction capacitance allows for high-frequency operation. The capacitance is variable, and has an inverse relation with reverse bias voltage. The resistor in parallel with the photodiode is the shunt resistance R_{SH}. If the shunt resistance is infinite, a current source can achieve ideal operation, and deliver all of its current to a load. The current-to-voltage ratio will be determined entirely by the load resistance. As the shunt resistance decreases to a comparable level to the value of the load resistance, it begins to impact on the current-to-voltage ratio, and the photodiode's noise increases. In real applications, shunt resistance is so high, up to tens or hundreds of megaohms, that it will not strongly influence overall performance. The series resistance is an equivalent resistance that includes the whole resistance from contacts, wire bonds, semiconductor material. This resistance tends to be quite low, several or a few tens of ohms, is usually not a major issue in the design of photodiode systems. However, excessive series resistance can reduce linearity.

In the analysis of frequency response of a TIA, we model the photodiode as a current source and a capacitor C_i. The capacitance across the input nodes of the op-amp including the capacitance of the photodiode and the internal capacitance of the op-amp, together with the feedback resistor forms a low-pass filter in the feedback path. Its frequency response can be characterized as the feedback factor:

$$\beta(j\omega) = \frac{1}{1 + j\omega R_f C_i} \tag{4.5}$$

Considering the effect of the low-pass filter, the response equation of the circuit is:

$$V_{out} = I_p \frac{-R_f}{1 + \frac{1}{A_{OL}\beta}} \tag{4.6}$$

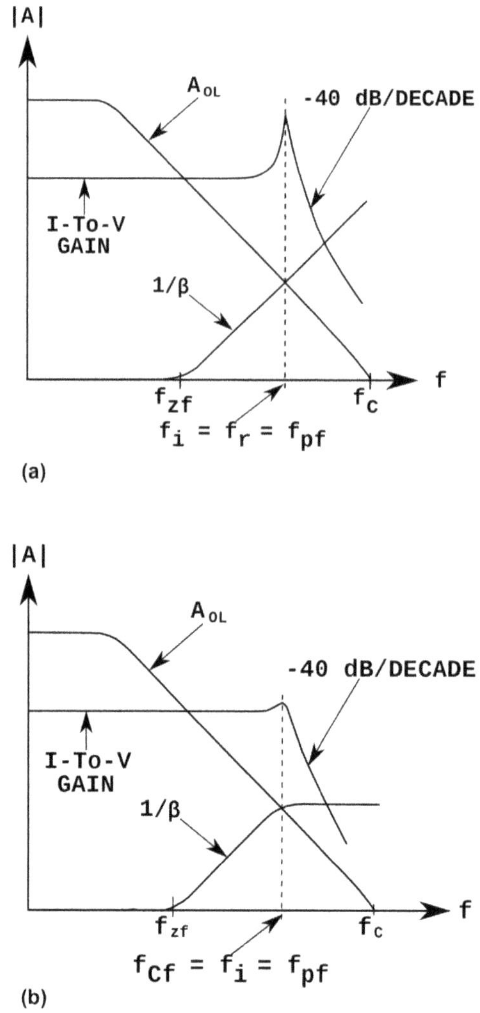

Figure 4.16. Bode plots of uncompensated (left) and compensated (right) TIA. These TIA Bode Plot 0 and 1 images in the center of the figure have been obtained and reproduced by the author from the Wikimedia website where they were made available by Zen-in under a CC BY-SA 3.0 licence. They are included within this article on that basis. They are attributed to Zen-in.

where A_{OL} is the open-loop gain of the op-amp. As shown in the left of figure 4.16, as long as $A_{OL}\beta$ is much greater than unity, the amplifier's response is close to the ideal condition at low frequencies such that $V_{out} = -I \times R_f$. At a higher frequency, there is a peak on the gain curve called gain peaking. It is a common problem when dealing with high-gain TIA, which induces an unstable behavior. Any noise signal can trigger an unstable circuit into oscillation. Therefore, the gain peaking is characterized by a ringing effect on the output voltage of the circuit. In order to maintain the stability of the circuit, the gain peaking needs to be eliminated by adding a compensated capacitor in parallel to the feedback resistor to create an additional pole in the response of loop gain, as shown in the right of figure 4.16. Besides, the gain peaking is reduced in this situation, producing a flatter overall response. The value of the compensated capacitor can be roughly determined according to the required gain-bandwidth product by:

$$C_{\mathrm{f}} \leqslant \frac{1}{2\pi R_{\mathrm{f}} f_{\mathrm{GBW}}} \tag{4.7}$$

where f_{GBW} is gain-bandwidth product. The required bandwidth of TIA is determined by the rise time of the returned pulse by the approximate relation that

$$\mathrm{BW} \approx \frac{0.35}{t_{\mathrm{r}}} \tag{4.8}$$

where t_{r} is the rise time of pulse measured from 10% to 90% amplitude of the rising edge. For a pulse with rise time of 1 ns, the minimum bandwidth is 350 MHz. It should be noted that the bandwidth must be high enough to preserve the shape of the pulse but should not be too much wider than needed, since the increasing bandwidth also brings more noise and incremental circuit cost. The more detailed calculation and frequency response analysis can be accomplished with the help of SPICE simulation tools like LTSpice.

A TIA PCB with an APD mounted on its surface is demonstrated in figure 4.17. At the right upper corner of the board is the APD. Its direction of emission is perpendicular to the surface of the board. The area at the downside close to the APD is the TIA. The TIA is built using an op-amp chip along with a feedback resistor and

Figure 4.17. PCB of TIA.

Figure 4.18. MAX 3658.

a feedback capacitor. The op-amp chip is OPA-657 from *Texas Instruments*, which is a widely used product with merits of low cost and ease-to-buy. To obtain high bandwidth, the gain of this TIA is not large. In this situation, the TIA mainly plays a role of converting current signal to voltage signal without providing much amplification. To further amplify the output voltage signal from the TIA, the TIA is followed by a second-stage amplifier to provide high gain for the signal and lift the signal to the required level.

A TIA built with individual components on PCB has a large footprint compared to an integrated chip. Moreover, the large stray capacitance induced by individual components, soldering, and PCB limits the operating bandwidth of a TIA. Therefore, in situations that require high speed or compactness, an integrated TIA chip is preferred. The chip shown in figure 4.18 for instance is MAX 3658 from *Maxim Integrated*, now a subsidiary of *Analog Devices*. MAX 3658 is an integrated TIA for receivers operating up to 580 MHz. It allows detection and amplification of extremely weak signal with short pulse width. With tiny footprint of 3 mm × 3 mm, it also features low noise, high gain and low power dissipation.

The power of returned pulses can be approximately determined by

$$P_{\mathrm{RX}} = \frac{G_{\mathrm{TX}} \times G_{\mathrm{RX}} \times P_{\mathrm{TX}} \times \pi R^2}{2\pi D^2} \tag{4.9}$$

where P_{TX} (P_{RX}) is the transmitted (received) power, G_{TX} (G_{RX}) is the efficiency of the transmitting (receiving) element (i.e., spectral responsivity), R is the radius of the optical receiving element, 2π is the solid angle at whose light is diffracted by the target, and D is the distance between the target and the sensing element. From the equation, the received power is inversely proportional to the square of the distance. This results in a wide dynamic range of the received power. When detection range from 1 to 100 m is needed, 40 dB power range is required for the receiver. If the minimum distance is down to 10 cm, then the dynamic range goes up to 60 dB. Such high dynamic range is a challenge to a TIA. Maximizing the SNR of the TIA increases the dynamic range of the system, which can result in improved target resolution, a lower variance in the target's absolute distance, and the ability to detect farther. Many researches focus on the improvement of TIA's dynamic range. For instance, Ma *et al* present 77 dB dynamic range low-power TIA with variable gain for LiDAR [14].

4.5 Timing discriminator

The ToF ranging method uses a counter to count the round-trip time between the emitter and the target. The counter is triggered by a digital logic signal by a timing discriminator. The ranging accuracy is directly determined by the timing accuracy. As we know, light travels at an extremely fast speed. As one ns passes, the light travels 30 cm. Nanosecond level error in time leads to large error in distance. Therefore, the discrimination of the arrival time of pulses highly dominates the timing accuracy. A timing discriminator is also referred to as a time pick-off circuit, which is vital in all timing systems. It provides an interface between the analog world of detectors and the digital world of logic systems. The discriminator handles analog pulses with varying shapes and amplitudes that arrives randomly in time. Analog input pulses that cross the timing discriminator threshold can be converted to standard logic pulses at the output of the timing discriminator. Achieving the arrival time of events with precision, accuracy and consistency is the primary task of a timing discriminator. An ideal timing discriminator produces a logic pulse at its output that is exactly related in time to the occurrence of an event. The three major factors that introduces errors to a timing discriminator are walk, drift and jitter.

Walk is the time movement of the output logic pulse from the timing discriminator relative to its input pulse, which is induced by changes in the shape and amplitude of the input pulse [15]. For two pulses with the same rise time and shape but different amplitudes, as shown in figure 4.19, if one fixed threshold level is set for the two pulses, the discriminated time for the two pulses is different. The output logic pulses differ in time with a walk error.

Drift is a long-term timing error that is induced by component aging and temperature variations in the timing discriminator circuits. Drift error is common in all kinds of sensors and instruments.

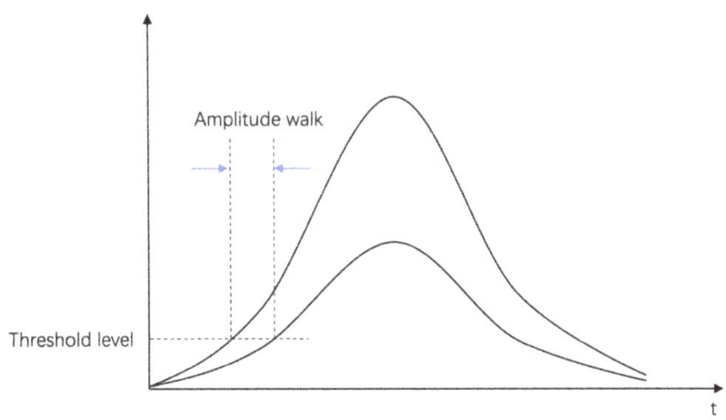

Figure 4.19. Walk error in amplitude.

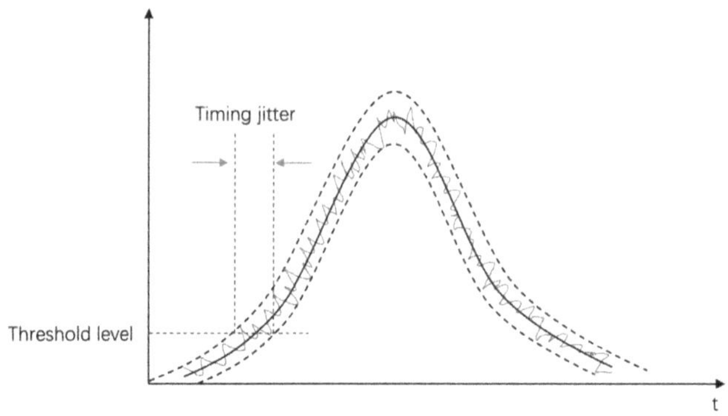

Figure 4.20. Timing jitter arising from random electronic noise.

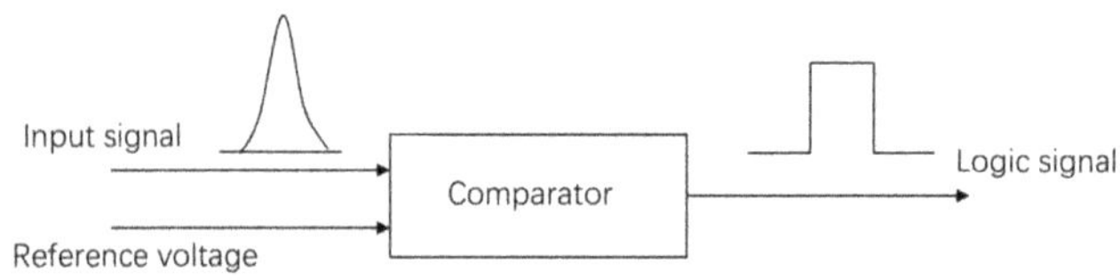

Figure 4.21. Schematic diagram of leading-edge discriminator.

Jitter is an usual but undesired error in electronics and telecommunications. It is the timing uncertainty caused by electronic noise or statistical fluctuations, which happens at the time that pulses cross the threshold, as illustrated in figure 4.20. The timing jitter has a positive relation with electronic noise

$$T_{\text{jitter}} = {e_n}\Big/\left(\frac{d_V}{d_t}\right) \tag{4.10}$$

where e_n is the voltage amplitude of electronic noise superimposed on the analog pulse, and d_V/d_t is the slope of the signal. Therefore, the timing jitter can be reduced by minimizing electronics noise. If the noise cannot be reduced, minimizing timing jitter can be achieved by setting the discriminator threshold at the point of maximum slope on the input analog pulse. Thus, it is best to preserve the fastest possible rise time for the input analog pulse (figure 4.20).

The decision is to turn on or off the counter at which moment of the emission and arrival of laser pulses directly affects the ranging accuracy. There are three main techniques to discriminate the timing of input analog pulses.

Leading-edge discriminator is a simple discriminating technique that incorporates a simple voltage comparator with its threshold set to the desired voltage, as illustrated in figure 4.21. It has a simple structure and is easy to build. The comparator generates a logic pulse when the leading edge of the analog pulse crosses this threshold, and the logic pulses end when the falling edge of the analog pulse crosses the threshold in the opposite direction. The initial transition of the logic pulse that corresponds to the threshold crossing on the leading edge of the analog

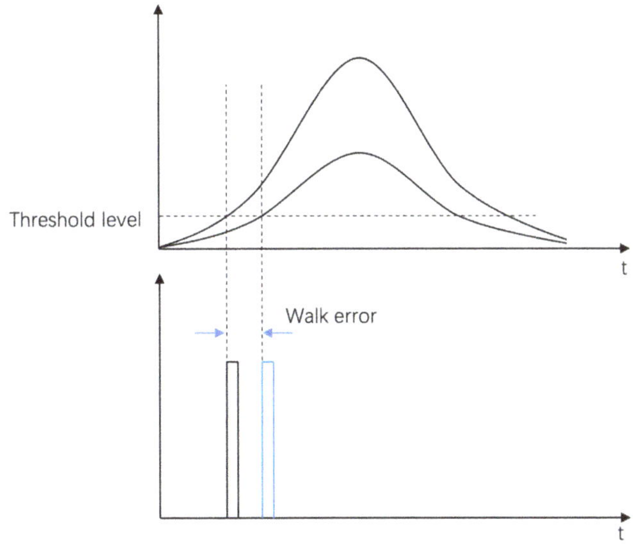

Figure 4.22. Walk error in leading-edge discriminator.

pulse is used to mark the arrival time of the analog pulse. If there is no electronic noise or amplitude changes, the arrival time of each analog pulse can be marked accurately with consistency by leading-edge discriminator. However, leading-edge discriminator is mainly limited by 'walk' error induced by the amplitude degradation of returned pulses. As shown in figure 4.22, with the same threshold level, the leading edges of two pulses with the same shape but different amplitudes reach the threshold at different times. This time difference is the walk error, which may result in large timing variations. Due to its simple structure, the leading-edge discriminator is normally used when optimum time accuracy is not essential.

Zero-crossing discriminator can greatly reduce the error induced by the amplitude walk, but at the cost of increased timing jitter. It requires that pulses have a bipolar shape as illustrated in figure 4.23. Even variations of amplitude exist, the time at which the waveforms cross zero volts is the same, and depends only on the amplifier shaping constants chosen to produce the pulses. The bipolar shape can be obtained by the derivation of the input pulses via a differentiator circuit made of either passive components or active op-amps. Therefore, the peak point of the original pulse is converted to the zero-crossing point. The zero-crossing point can be detected using an op-amp. It is basically a voltage comparator whose output changes when the input signal crosses the zero of the reference voltage level. For the topology of zero-crossing detector shown in figure 4.24, the zero-crossing detector is built on the basis of op-amp 741 IC for instance. The zero-crossing detector changes the op-amp's output state when the input signal crosses the zero-reference voltage. This is done by grounding the op-amp non-inverting input as the zero-reference voltage and applying the attenuated input to the inverting input. When the input bipolar signal passes through zero and goes in a positive direction, the output voltage is driven into

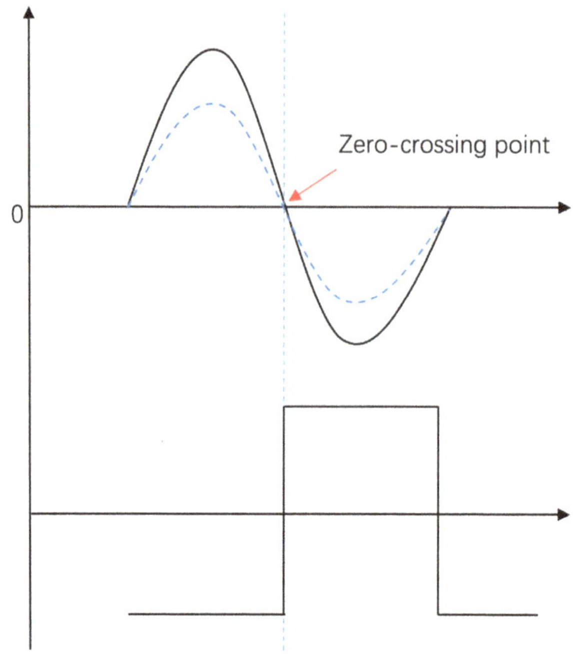

Figure 4.23. Principle of zero-crossing discriminator.

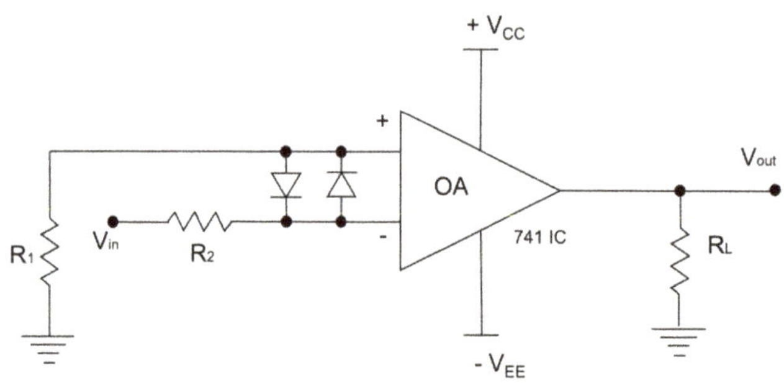

Figure 4.24. Topology of zero-crossing discriminator.

negative saturation. Similarly, when the input voltage passes through zero and goes in the negative direction, the output voltage is driven to positive saturation. The two diodes connected between the inverting and non-inverting terminals are clamp diodes to protect the op-amp from damage due to increase in input voltage. The differential input voltages are clamped to either +0.7 V or −0.7 V. Due to the same shape of returned pulses, the arrival time of peak point is consistent, even in the presence of amplitude variations. Therefore, the walk error induced by amplitude variations can be eliminated by the zero-crossing discriminator. However, the amplitude degradation and additional noise are brought by the shaping stage of the differentiator, making the zero-crossing point susceptible to statistical

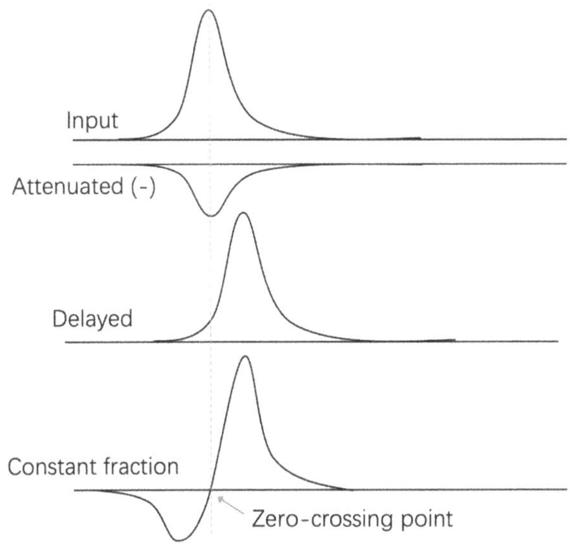

Figure 4.25. Principle of constant fraction discriminating.

fluctuations. In the situation where the amplitude of input pulse is small, empirically less than 10%–15% of the maximum pulse amplitude, a leading-edge discriminator outperforms a zero-crossing discriminator.

Constant fraction discriminator essentially eliminates amplitude-dependent walk error in time for signals having consistent rise times, making it always trigger at the optimum fraction of the pulse height for any pulse height. Its net result is optimum time resolution over a wide dynamic range of pulse heights. Pulse shaping is employed in constant fraction discriminator as its principle illustrated in figure 4.25. The input signal is split into two parts such that one is attenuated to a fraction ($f < 1$) of the original amplitude and then inverted, and the other one is delayed by a fixed time interval. The two signals after conversion are subsequently added to generate the constant fraction timing signal. The delay is used to align the optimum fraction point at the leading edge of the delayed pulse with the peak amplitude of the attenuated pulse. The delayed time t_d can be calculated by

$$T_d = T_r(1 - k) \tag{4.11}$$

where T_r is the rise time of input pulse and k denotes the attenuation fraction of attenuator. The attenuation fraction should be in the range of 0.2–0.5, it may cause a great standard deviation of the time discriminator otherwise. Consequently, adding the delayed and attenuated pulses yields a bipolar signal with a zero-crossing point that corresponds to the original point of optimum fraction at the delayed signal.

As the system block of constant fraction discriminator illustrated in figure 4.26, the constant fraction discriminator incorporates a zero-crossing discriminator elaborated in the above to detect the zero-crossing point of the bipolar signal and then produce a logic pulse, thus providing a time marker at the optimum fraction of pulse height. The attenuator circuit can be accomplished by resistance voltage

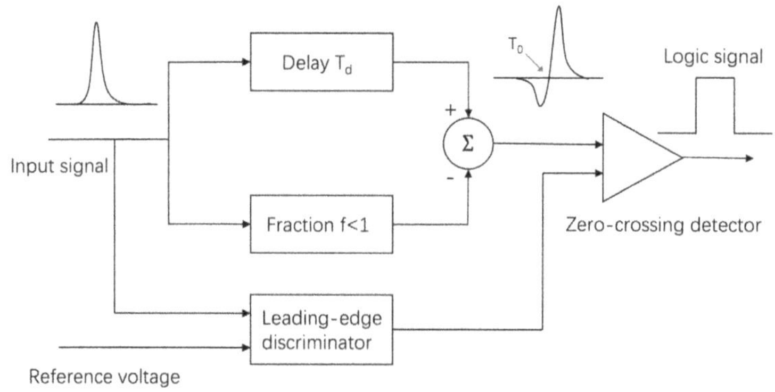

Figure 4.26. Schematic diagram of constant fraction discriminator.

divider. The delay circuit employs a first-order *RC* delay circuit, an *LC* delay circuit or a dedicated delay-line circuit. It is necessary to be noted that the inverted conversion can be deployed after either delay line or attenuator, depending on the phase of input signal. Walk and jitter can be minimized by proper adjustment of the zero-crossing reference, and by selection of the correct attenuation fraction and delay. Since the time discrimination of zero-crossing is independent of pulse amplitude, the constant fraction discriminator characterizes almost zero walk. Parallel to the constant fraction discriminator, a leading-edge discriminator provides energy selection capability and prevents the sensitive zero-crossing device from triggering on the noise on the constant fraction baseline. The input signal is split to two channels into constant fraction discriminator and leading-edge discriminator, respectively. The leading-edge discriminator produces triggering pulse when the input pulse crosses a constant level, which is feasible to avoid false trigger resulted from noise.

The amplitude and slope of the differential of attenuated and delayed inputs are dependent on the amplitude of the input signal. This dependency leads to the variation of charging time or response time of the comparator, causing time walk error in constant fraction discriminator. As the analysis on the dynamic range of the photodiode output along with the TIA output, the dynamic range of signal power fed into time discriminator is also very wide. This large variation of input amplitude may cause time walk error. To solve the above problem, an automatic gain control (AGC) is adopted before a constant fraction discriminator to reduce the dynamic range of input signal fed into the discriminator [16]. Hence, variations in the amplitude and slope of the differential input of the comparator are reduced, so too the variation in the response time of the comparator. AGC is a close-loop feedback regulating circuit in an amplifier or chain of amplifiers, the purpose of which is to maintain a suitable signal amplitude as its output, despite variation of the amplitude of the input signal [17]. The peak level of the input signal is used to dynamically adjust the gain of the amplifier, enabling the circuit to work satisfactorily with a greater range of input signal levels. An AGC usually consists of two parts: a peak detector circuit and a variable gain amplifier (VGA), as illustrated in figure 4.27. Its operating mechanism is that the input signal to be gain controlled goes to a diode

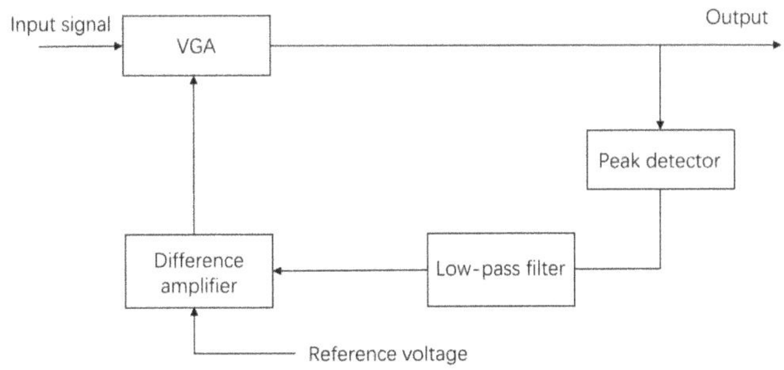

Figure 4.27. Block diagram of AGC.

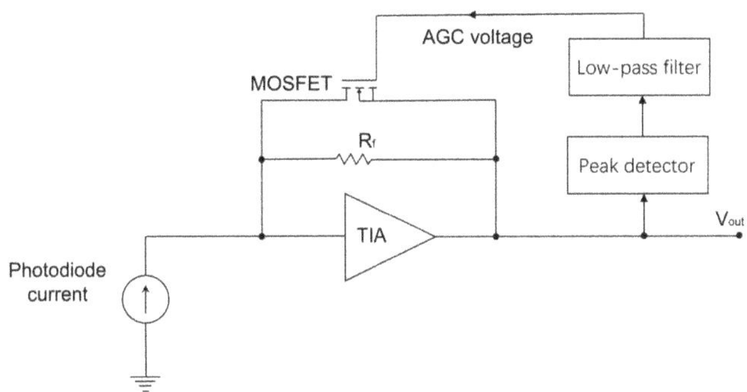

Figure 4.28. Schematic diagram of TIA with AGC.

and capacitor, which produce a peak-following DC voltage to be fed to gain blocks to alter their bias together with gain [18]. The input signal passes through the VGA to produce the output level to be stabilized. The peak detector's output is compared against reference voltage to produce a differential voltage for the gain control of the VGA. Between the VGA and the peak detector, an attenuator or amplifier can be added to align the maximum output level of the VGA with the maximum input level of the detector. When the input signal has lower amplitude, the VGA has a large gain to amplify the input signal to a higher level. While the VGA provides little amplification on higher input signal. Hence, the wide range of input signal can be reduced to a narrow range to eliminate walk error of discriminator. VGA is an electronic amplifier that varies its gain depending on a control voltage, which can be implemented using analog control VGA, programmable gain amplifier, and cascaded attenuators with selection mechanism. Analog control VGA provides continuous gain control over a wide dynamic range for a variety of frequency bands. Programmable gain amplifier can provide excellent gain accuracy. The output of peak detector needs to be converted to digital signal using an analog-to-digital converter to encode the peak level. The encoded signal with multiple digits is used to control the programmable gain amplifier. AGC can be incorporated with TIA to automatically control the feedback resistance of TIA to extend the dynamic range of TIA, as illustrated in figure 4.28. In a typical AGC, the scheme of peak detector after VGA cannot handle the abrupt change of input signal because the peak is detected after the amplification of the input signal. To control the gain of the input signal in 'real time', the peak detector is followed by the VGA and the input signal is split into two channels. One is for peak detection and the other is delayed. After the peak is detected and the suitable gain of VGA is set, the delayed input signal is fed into the VGA. But for signal with wide pulse width, very large delay is not easy to realize.

To realize automatic control of gain on input signal, its amplitude level is required to control the gain of the amplifier. The amplitude peak level can be obtained by peak detector circuit. Peak detector is a circuit that is able to determine the peak value of an input signal in a rapidly changing waveform and then stores the peak value. Figure 4.29 is a basic peak detector circuit consisted by diode and capacitor. In the positive phase of signal, diode D is forward biased and capacitor C starts charging. When the input reaches its peak value, the capacitor gets charged to positive peak value. As the input decreases in negative phase, the diode D is reversely biased, and the capacitor is isolated and holds the peak value of the previous phase. By the time of moving further, if the circuit detects a higher peak, the new peak value is stored by the capacitor until it is discharged. As the input voltage reduces below the values of the capacitor voltage, it causes the diode to get reversely biased. In such condition, the capacitor retains the value until the input again exceeds the value stored in the capacitor. Hence, the peak level is detected by the peak detector circuit.

However, practically, output is taken across some load resistor, so when the input voltage decreases, the capacitor discharges through the load. The value of the load needs to be very large so that the capacitor discharges very slowly to hold the charge. To avoid the discharging through the load, an op-amp is placed between the input and the diode, as illustrated in figure 4.30. In the positive phase, the output of op-amp is positive, so the diode is forward biased and the capacitor charges to the peak

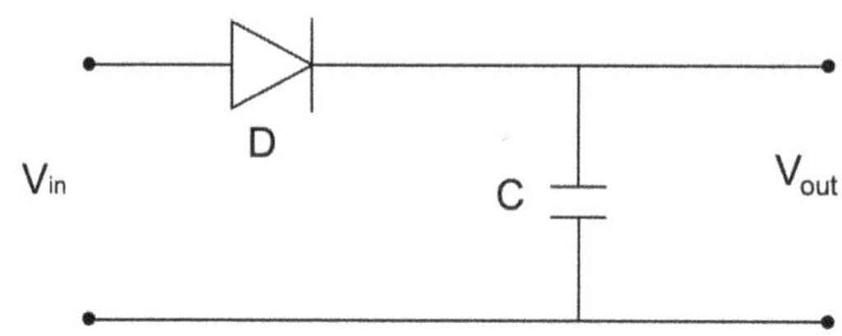

Figure 4.29. Schematic diagram of basic peak detector.

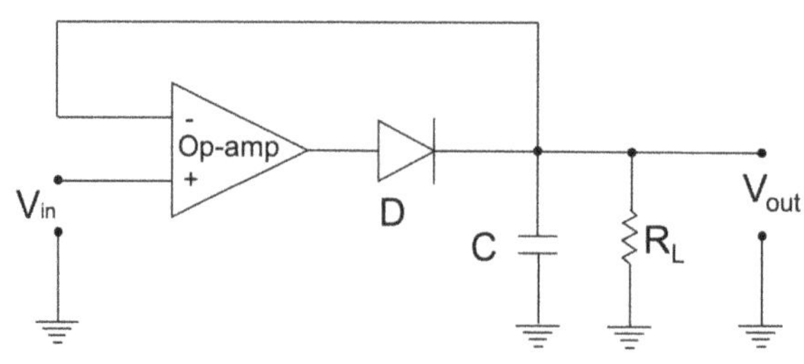

Figure 4.30. Schematic diagram of op-amp-based peak detector.

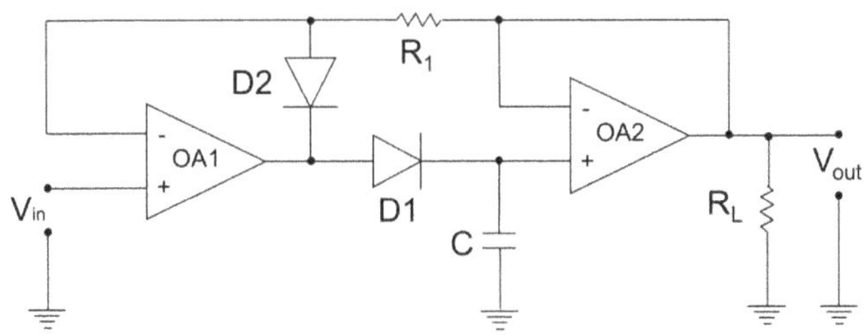

Figure 4.31. Schematic diagram of improved op-amp-based peak detector.

value of input signal. In the negative phase, when input decreases, the diode is reversely biased and the capacitor is isolated and holds the charge of previous phase. Since the diode is reversely biased, the op-amp is in open-loop condition and goes into saturation. The capacitor still discharges rapidly through the load, if the load value is not large enough. Hence, an improved op-amp-based peak detector with a second op-amp acts as a voltage follower is proposed, as illustrated in figure 4.31. The second op-amp OA2 acts as a voltage follower, of which input impedance is very high. So, the capacitor discharges very slowly, it almost holds the charge. The output voltage is nothing but voltage across the capacitor.

4.6 TDC and timing

Timing is essential for ToF LiDAR distance measurement, its precision is critical to the LiDAR's ranging precision. To obtain 1 cm accuracy, the accuracy of the time interval measurement should be 67 ps. As elaborated in the above section, the emitted and arrival time of pulses can be accomplished using timing discriminator. Since the time of emitted and arrival are discriminated, their interval needs to be measured by a counter. A time-to-digital converter (TDC) is used to measure the time interval between the emission and arrival of laser pulses to get the correct timing [19]. TDC is a chip that counts the time between start and finish signals and reverts it to the measurable digital format. The timing discriminator generates logic-level timing pulse to trigger TDC to count. It can be considered as a combination of a time-analog and analog-to-digital converter to measure time intervals down to picosecond range and output time information in digits [20].

There are two main implementation techniques: ASIC-based TDC and FPGA-based TDC. ASIC-based TDC holds advantages of low cost, mass production, ease-to-buy, and compact size, while the primary merit of FPGA-based TDC is its reconfigurability and integration with other digital logics. ASIC ones of TDC 7201 from *Texas Instruments* provide 55 ps resolution, and GP-22 from *ACAM* provides 90 ps resolution, 45 ps using double resolution mode and 22 ps using quadruple resolution mode. With delicate design, FPGA-based TDC can achieve sub-ps resolution using a high-end FPGA chip.

A traditional approach to implement TDC is first to convert time interval into voltage using an integrator, a capacitor for instance, and then digitalize the voltage by sampling using an analog-to-digital converter [21]. The precision of TDC based on this approach relies on the resolution in bits of the analog-to-digital converter.

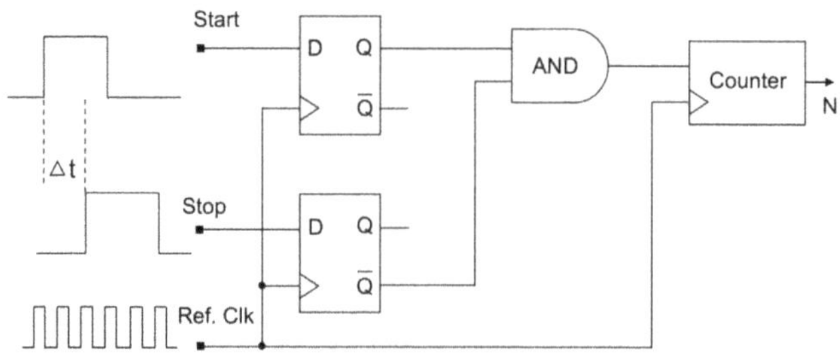

Figure 4.32. Schematic diagram of single-counter TDC.

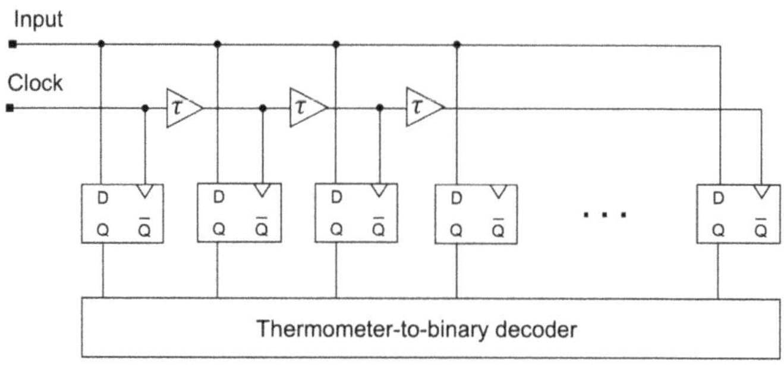

Figure 4.33. Schematic diagram of delay-line TDC.

The linearity of the integrator has influence on the error of timing. This traditional approach is also considered as an analog one, and very sensitive to environmental changes. To hold the scaling and robustness arguments, digital conversion approaches are widely used. The simplest technique to quantize a time interval is to count the cycles of a high-frequency reference clock between a start and a stop pulse, as illustrated in figure 4.32. The arrival of *Start* and *Stop* pulses is discriminated by D-type flip-flops. The *AND* gate ensures that the counter is enabled only when *Start* and *Stop* are logically different. The resolution of this approach is constrained by the frequency of the reference clock, which can be no higher than a single clock period. Typically, a TDC requires a reference clock with higher frequency and good long-term stability, of which timing resolution can be increased by a higher clock frequency. However, the higher the clock frequency, the higher the power consumption for the generation and the processing of the clock signal. More importantly, it is not easy to realize very fast on-chip clock oscillator.

To increase the timing resolution beyond the maximum feasible clock frequency, the reference clock is delayed by a chain of digital delay elements. Hence, the resolution depends on the delay time of delay elements in the chain. The operating principle of TDC based on delay line is illustrated in figure 4.33. The elements that compare the input signal to the reference clock can be latches or D-type flip-flops. Each enabling signal of D-type flip-flop is delayed by a fixed delay τ. To ensure that τ is accurate with high consistency and stability, the delay chain can be implemented and stabilized by a delay-locked loop (DLL). The thermometer-to-binary decoder detects bit changes of raw data between '0' and '1', and converts data from the thermometer code into binary code, making it readable for processors. The drawback to this implementation is that the temporal resolution cannot be higher than the delay through a single gate in the semiconductor technology used. Since it can be easily constructed with solely digital components using any standard CMOS process, delay-line TDC is well suited for both ASCI- and FPGA-based TDC, making it widely used in timing measurement systems. It is capable of performing a measurement on every clock cycle and can be operated at relatively higher speed.

To achieve sub-gate temporal resolution, the delay-line converter can be constructed with a Vernier delay line. This approach provides higher resolution of the difference between two delays. Two individual DLLs should be implemented for each delay chain to make them reasonably accurate. With higher resolution, the structure of this approach is more complicated than a typical delay-line TDC.

References

[1] https://en.wikipedia.org/wiki/Photodiode (accessed in October 2021)
[2] Quimby R S 2006 *Photonics and Lasers: An Introduction* 1st edn (New York: Wiley)
[3] Pearton S, Ren F and Mastro M 2018 *Gallium Oxide: Technology, Devices and Applications* 1st edn (Amsterdam: Elsevier)
[4] https://rp-photonics.com/dark_current.html (accessed in October 2021)
[5] Yotter R A and Wilson D M 2003 A review of photodetectors for sensing light-emitting reporters in biological systems *IEEE Sen. J.* **3** 288–303
[6] https://en.wikipedia.org/wiki/PIN_diode (accessed in October 2021)
[7] Liu J, Cristoloveanu S and Wan J 2021 A review on the recent progress of silicon-on-insulator based photodetectors *Phys. Status Solidi* A **21824** 2000751
[8] Campbell J C 2015 Recent advances in avalanche photodiodes *J. Lightw. Technol.* **10** 777–87
[9] https://rp-photonics.com/avalanche_photodiodes.html (accessed in October 2021)
[10] Behroozpour B, Sandborn P A M, Wu M C and Boser B E 2017 Lidar system architectures and circuits *IEEE Commun. Mag.* **55** 135–42
[11] Chen G, Wiede C and Kokozinski R 2021 Data processing approaches on SPAD-based d-TOF LiDAR systems: a review *IEEE Sens. J.* **21** 5656–67
[12] Robbins M F 2021 *Ultimate Electronics: Practical Circuits Design and Analysis* ch 7 https://ultimateelectronicsbook.com/ (El Paso, TX: CircuitLab) (accessed in November 2021)
[13] https://en.wikipedia.org/wiki/Transimpedance_amplifier (accessed in November 2021)
[14] Ma R, Liu M, Zheng H and Zhu Z 2018 A 77-dB dynamic range low-power variable-gain transimpedance amplifier for linear LADAR *IEEE Trans. Circuits Syst. II* **65** 171–5
[15] https://ortec-online.com/-/media/ametekortec/other/fast-timing-discriminator-introduction.pdf (accessed in November 2021)
[16] Wang H, Li X, Xu L and Xu T 2017 LiDAR ranging system based on automatic gain control and timing discriminators *2017 IEEE Int. Instrumentation and Measurement Technology Conf. (I2MTC)(May 2017) (Turin, Italy)*
[17] https://en.wikipedia.org/wiki/Automatic_gain_control (accessed in November 2021)
[18] Lim H 2014 Constant fraction discriminator involving automatic gain control to reduce time walk *IEEE Trans. Nucl. Sci.* **62** 2351–6
[19] https://en.wikipedia.org/wiki/Time-to-digital_converter (accessed in November 2021)

[20] Roberts G W and Ali-Bakhshian M 2010 A brief introduction to time-to-digital and digital-to-time converters *IEEE Trans. Circuits Syst. II* **57** 153–7
[21] Henzler S 2010 *Time-to-Digital Converters, Springer Series in Advanced Microelectronics* 1st edn (Dordrecht: Springer)

Chapter 5

Scanner

This Velodyne High-Def LIDAR image has been obtained by the author from the Wikimedia website https://commons.wikimedia.org/wiki/File:Velodyne_High-Def_LIDAR.jpg where it was made available by Steve Jurveston under a CC BY 2.0 licence. It is included within this article on that basis. It is attributed to Steve Jurveston.

A complete functional LiDAR consists of a transmitter, a receiver, and a scanner. The former two modules have been elaborated in the last two chapters. Without a scanner, the combination of a transmitter and a receiver is referred as a rangefinder or 1D LiDAR that performs point-to-point distance measurement. A scanner is the key module that directly determines field of view (FOV), angular resolution (spatial

doi:10.1088/978-0-7503-3723-6ch5

resolution), and scanning frequency. With 1D rotating in horizontal direction, a LiDAR is referred as a 2D LiDAR that scans along a 2D plane. While a 3D LiDAR scans all the 3D space via steering simultaneously in both horizontal and vertical directions. An excellent scanner empowers outstanding the spatial awareness capability of a LiDAR. In this chapter, different scanning mechanisms including mechanical, micro-electro-mechanical systems (MEMS)-based, and optical phased array (OPA)-based will be elaborated.

5.1 Scanner specifications

FOV is one of LiDAR's primary specifications, indicating the span of area that can be detected by a LiDAR. A scanner is what expands the detectable FOV of a LiDAR, in the absence of which a LiDAR is a point-to-point rangefinder. The FOV of a LiDAR can be either a 2D horizontal plane or a 3D space in both horizontal and vertical planes, as illustrated in figure 5.1. The FOV is typically represented in degrees, and can be calculated in horizontal and vertical directions by

$$\mathrm{FOV_H} = \Delta\theta = \theta_2 - \theta_1 \tag{5.1}$$

$$\mathrm{FOV_V} = \Delta\varphi = \varphi_2 - \varphi_1 \tag{5.2}$$

where θ_1 is the minimum azimuth angle, θ_2 is the maximum azimuth angle, φ_1 is the minimum elevation angle, and φ_2 is the maximum angle.

Angular resolution describes the ability of a LiDAR to distinguish small details of an object, thereby making it a major specification of a LiDAR. For a mechanical LiDAR, the minimum step of a servo or step motor determines the minimum angular resolution of the LiDAR. A 2D LiDAR has only angular resolution in the horizontal plane. If the size of an object in a far position is smaller than the corresponding length to the minimum angular resolution of a 2D LiDAR, it will be missed by the scanning of the LiDAR. A 3D LiDAR has both horizontal and vertical angular resolutions. Compared to its horizontal angular resolution, the vertical angular resolution of a 3D LiDAR possesses a more important position. The resolution of a 3D LiDAR is usually characterized by the quantity of beams, which can be calculated by

$$N = \frac{\mathrm{FOV_V}}{d\varphi} \tag{5.3}$$

where $d\varphi$ is angular resolution. A 3D LiDAR with more than 32 beams in vertical direction can be categorized as a high-resolution LiDAR (not an official standard). Except for the design of scanning pattern, the quantity of beams also can be multiplied by increasing pairs of transmitters and receivers. For mechanical rotating LiDARs that scan only in horizontal direction, their beams are only increased by incremental transceiver channels.

The FOV determines the detectable span area of a LiDAR, and its ability of detecting small-sized objects is determined by the angular resolution. The ability of detecting changes of fast-moving objects relies on the refresh per complete span area,

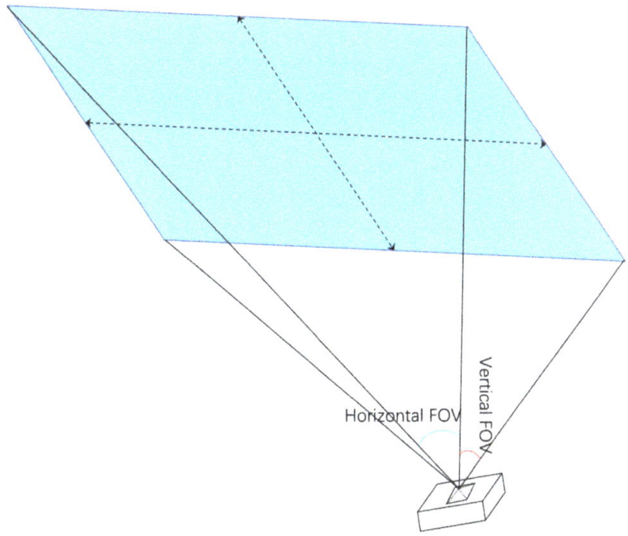

Figure 5.1. Schematic diagram of FOV.

i.e., frame rate or scanning frequency. Its unit is frame rate per second (FPS) or Hz. Here, the concept of frame rate is similar to that of a camera, of which the difference is that the frame for a LiDAR is 3D information rather than color information for a camera. For a 2D LiDAR, the definition of frame rate is not applicable. The frequency in terms of Hz is used to characterize the scanning speed of a 2D LiDAR, which usually denotes the quantity of turns it can scan within one second. In a line-by-line scanning manner, the spatial resolution is contradictory to the frame rate. The rotating speed has its intrinsic physical limit for a motor. More scanning lines in a span area per second lead to fewer completed scans over the span area. For application in autonomous driving, the requirement of scanning speed is relatively higher of 10–20 FPS (Hz).

5.2 Optical configuration

The arrangement of transmitter and receiver has two schemes: bistatic and monostatic, as illustrated in figure 5.2. Bistatic LiDAR is a LiDAR system comprising a transmitter and receiver that are separated by a distance comparable to the expected target distance [1]. The optical path of a transmitted laser is not identical or parallel to the path of the returned laser. Conversely, the transmitter and receiver in a monostatic LiDAR are co-located and their optical paths are parallel. Due to the large separated distance between the transmitter and receiver, bistatic LiDAR is not able to detect a large distance range. Besides, the bistatic scheme possesses a large footprint compared to the monostatic scheme, which is not suitable for autonomous driving, robotics, and consumer electronics applications. Therefore, most LiDARs adopt the monostatic scheme.

The optical path of a transceiver can be categorized as coaxial and biaxial, as illustrated in figure 5.3. In the coaxial arrangement, the axis of emitted laser beam is

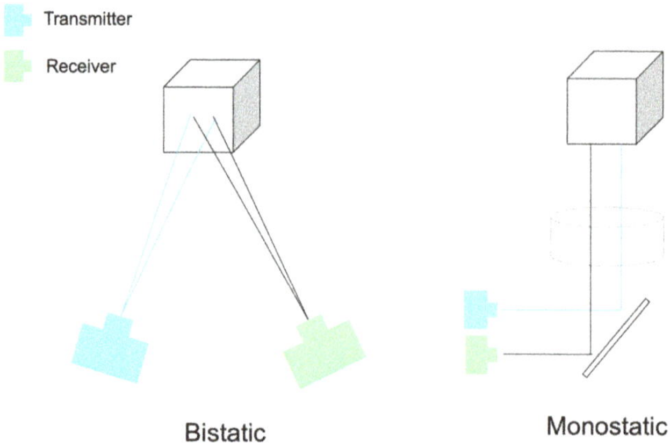

Figure 5.2. Schematic diagram of bistatic and monostatic LiDAR.

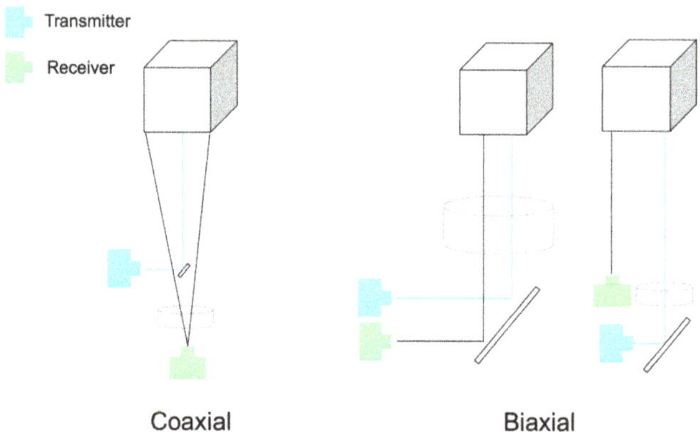

Figure 5.3. Schematic diagram of coaxial and biaxial schemes.

exactly coincident with the axis of photodetector optics. Therefore, the photodetector can 'see' the laser beam since the position is very close to the zero range. The near-field backscattering problem that may saturate the photodetector in a coaxial system can be overcome by either gating the photodetector or by use of a fast shutter or chopper to block the near-field scattering. There are lots of LiDARs using the biaxial scheme. In the biaxial scheme, the emitted laser beam only enters the FOV of the receiver optics beyond some predetermined range. So, there exists a blind range near the LiDAR, where the laser beam cannot be collected by the receiver. Based on that, the biaxial scheme helps avoiding near-field backscattered radiation that may saturate the photodetector. The choice of biaxial or coaxial arrangement is usually determined by the detection range and scanning component. If near-field range is desired, a coaxial scheme is preferred as it provides full overlap of the receiver's FOV with the emitted laser beam. If near-field range is not desired, the biaxial scheme may help prevent the saturation of the photodetector by strong near-field scattering. Scanning capability can also come into play for the selection of biaxial or coaxial.

5.3 Scanning schemes

Scanning schemes of LiDAR can be roughly categorized into mechanical, hybrid solid-state, and solid-state, as shown in figure 5.4. The development of scanning techniques moves from mechanical to hybrid solid-state and then to solid-state. There are no rigorous boundaries between each of the two schemes. As a traditional and popular scanning scheme, entirety rotating that employs a motor to drive a scanning head is a mechanical LiDAR. However, for a mirror rotating scheme that is categorized into hybrid solid-state here, it is considered as a mechanical scheme sometimes. In some technical notes of LiDAR or automobile manufacturers, MEMS-based micromirror is classified as solid-state, probably because of the micro-vibration of a tiny mirror compared to rotating large mirrors or scanning heads. The solid-state schemes normally include optical phased array and flash techniques. With low power consumption, compactness and low cost, the solid-state schemes are considered to be ultimate solutions for autonomous vehicles, robotics and consumer electronics.

5.4 Mechanical rotating

Mechanical rotating is a traditional yet simple manner to realize the beam steering of a LiDAR. Velodyne, who developed the first 3D LiDAR in the industry, adopts this technique and has continued to launch new rotating products to increase the vertical resolution. The merits of mechanical rotating are low cost, 360° FOV, and low complexity. It is the only way to achieve 360° scanning among all the scanning mechanisms (mirror rotating technique scans close to 360° with a small blind area). For a 2D LiDAR, a pair of transmitter and receiver is mounted on a motor. The

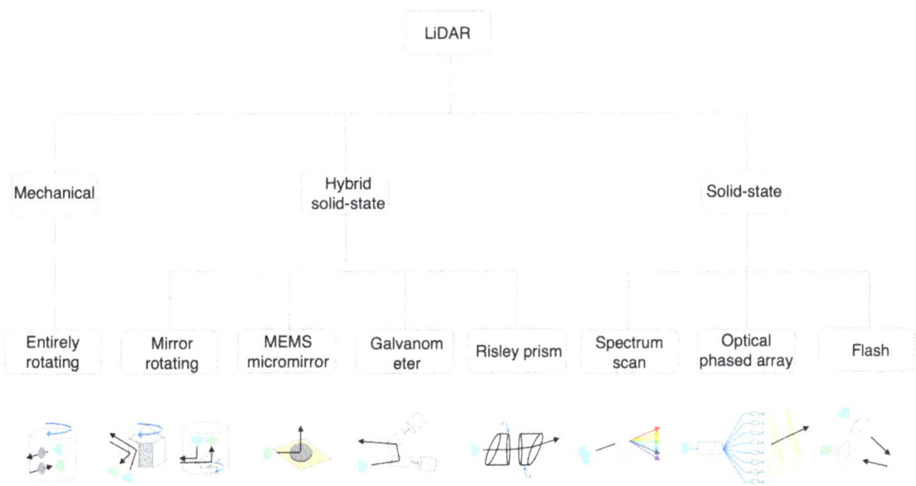

Figure 5.4. Scanning schemes of LiDAR.

motor that rotates along one direction in the horizontal drives the transceiver to rotate synchronously, as illustrated in figure 5.5. The minimum step or rotating speed of the motor determines the angular resolution and scanning frequency of the 2D LiDAR. The position and spun angle are obtained from an encoder inside the motor. The angular is fixed and unchangeable, but the spatial resolution varies with the distance of object. Based on the triangular relation, the spatial resolution gets larger with the increasing distance. Since the transceiver rotates in 360°, connecting wires transmitting power and data between the transceiver and mother board will twist and damage the LiDAR. Hence, a slip ring is required to transmit power and data. Apart from a slip ring, data can also be transmitted using wireless optical communications between the transceiver board and mother board. It is a wireless way to communicate using an LED as transmitter and a PIN as receiver. The power can also be transmitted in a wireless manner by magnetic field via inductive coupling between coils. Thus, the twisting issue of wires can be solved and the LiDAR scans 360° surrounding. Due to their 1D scanning, 2D LiDARs are not suitable for high-level autonomous driving tasks. They can be only used as gap filler in blind areas for driving assistance. Besides, the applications of 2D LiDARs involve anti-collision and navigation of service robots due to their low cost and small size.

The extended scanning in vertical direction for a 2D LiDAR can be implemented by another motor that rotates in a direction perpendicular to the horizontal one. Actually, in practice, the vertical scanning is accomplished by adding more pairs of transmitters and receivers aligned with fixed tilted angle in vertical, as illustrated in figure 5.6. The angular resolution in vertical is directly determined by the quantity of transceivers. A typical configuration has a vertical array of 16 lasers scanning 30° giving a vertical resolution of 2°. Compared to adding another motor, the approach of incremental transceivers brings much richer points and denser coverage. Because a pulse laser has a physical limit in emitting laser pulses per second, the extended scanning in vertical reduces the density in horizontal. For multi-channel transceivers, emitted laser pulses per second are multiplied by the quantity of transceivers, resulting in much denser points. However, transceivers cannot be added without limit due to the size and cost. What is more important is that the optical alignment becomes challenging with the increasing transceivers, hindering the increase of spatial resolution for a mechanical rotating LiDAR. The period of assembly and test

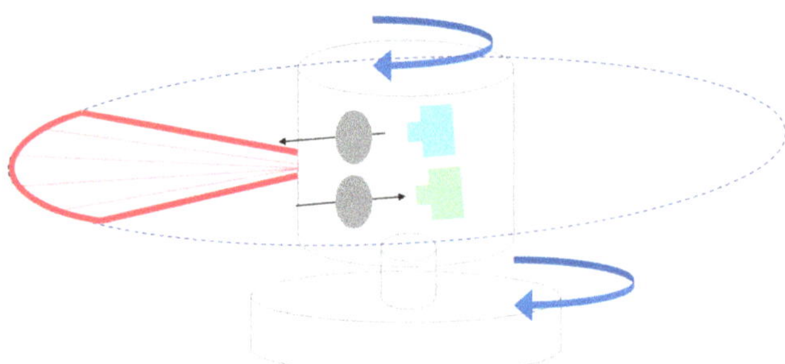

Figure 5.5. Schematic diagram of mechanical rotating LiDAR.

is also extended, which further pushes up the cost. Except for its bulky size, the price of a mechanical *Velodyne* LiDAR with 64 beams is even higher than a luxury sedan. Nowadays, the highest resolution for a mechanical LiDAR is 128 beams, first achieved by top LiDAR manufacturer *Velodyne*.

5.5 Mirror rotating

The direction of laser beams can be changed by either rotating the laser source itself or with an optical system of rotating mirrors. Rotating the scanning head is a straight way to 'look' around, like turning one's head. However, the scanning head that contains transmitters, receivers, lens and electronic boards is usually heavy, making it rotate at a slow rate [2]. We know that a light beam will be deflected by a reflective surface. Therefore, the scanning of a LiDAR can also be realized by hitting laser pulses on a mirror and then rotating the mirror to steer the direction of the laser. The mirror in the mirror rotating approach is much lighter compared to the entirety rotating approach, allowing faster rotating speed with greater accuracy. As a 2D LiDAR demonstrated in figure 5.7 shows, the mirror is 45° tilted inside a

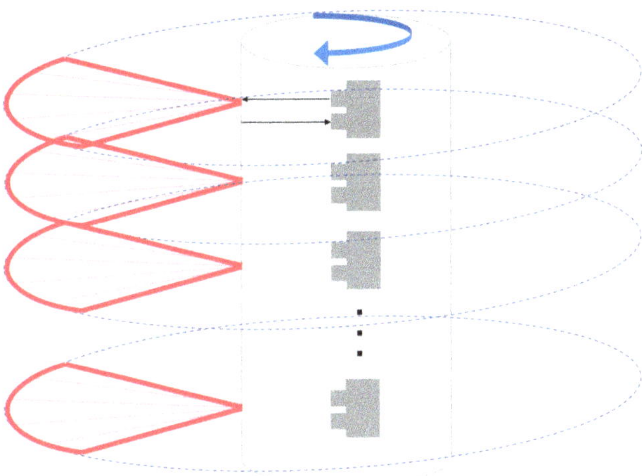

Figure 5.6. Schematic diagram of multi-channel transceivers.

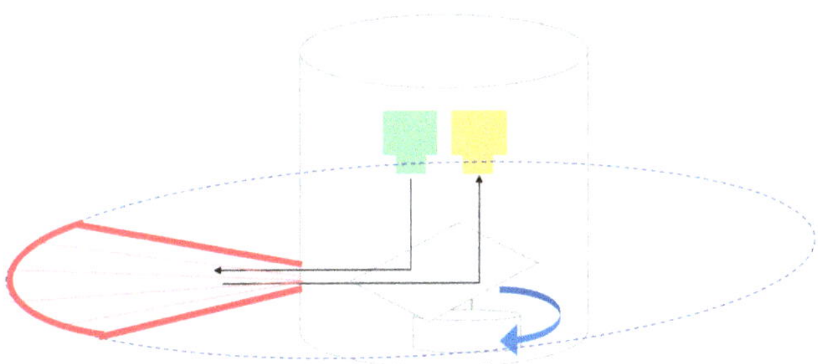

Figure 5.7. Schematic diagram of mirror rotating 2D LiDAR.

LiDAR. The transmitter and receiver are both on the top of the LiDAR pointing at the surface of the mirror. Also, the transceiver can be mounted at the bottom of the LiDAR depending on its design scheme. The emission and receiving laser beams share the same path. After perpendicular deflection, laser beams scan horizontally with the mirror rotating. This is a typical scanning scheme that was adopted by *SICK* 2D LiDARs, which is a global manufacturer of sensors and sensor solutions provider for industrial applications based in Waldkirch, Germany. By using more transceiver channels tilted at different angles, multi-beams scanning can be achieved. ALASCA LiDAR from *Ibeo* performs four beams scanning using the above scheme [3]. *Ibeo* is an automobile sensor manufacturer based in Hamburg, Germany.

Ibeo's SCALA LiDAR also uses four transceiver channels to perform four beams scanning, but employs a polygon mirror with surfaces pointed in exactly the same direction. By changing the pointing angles of each mirror, multi-beams scanning can be implemented even with only one transceiver channel. Polygon mirror is a rotating component with a series of flat reflective surfaces around the perimeter that is used in scanning systems to reflect light from the source onto the object being scanned, as demonstrated in figure 5.8. The motion of a polygon mirror is a simple rotation. Since a polygon mirror has multiple surfaces, even slow rotating can scan the angle rapidly. A polygon mirror only scans in one dimension, but it is possible to have surfaces tilted at different angles, providing an elevation step scan [4]. For instance, a hexagon mirror that has 6 surfaces scans over 120°. The mechanical scan is 60°, but the optical scan angle is 120°. The beam-steering angle is always twice the mechanical rotation angle as angle of incidence equals angle of reflection. The more the surfaces on the polygon mirror, the smaller the scan angle before the scan jumps to the next scan, but the physically larger the polygon. A polygon mirror always scans along unidirection with no fly-back period. As soon as the beam crosses from one flat surface to the next, the next scan begins from the original angle. Scan loss happens when the beam hits the corners between two surfaces. To avoid this issue, measurements in the vicinity of corners can be neglected by algorithm postprocessing. A drawback of the polygon mirror is its size, especially for polygons

Figure 5.8. Polygon mirror. The rotating polygonal mirror board image on the right has been obtained by the author from the Wikimedia website where it was made available by Raimond Spekking under a CC BY-SA 4.0 licence. It is included within this article on that basis. It is attributed to Raimond Spekking.

with more surfaces. Another drawback of the polygon mirror is that the center of rotation is not on the mirror surface. The center of rotation is in a different location to the surface of polygon, the steered beam will translate back and forth slightly, causing distortion at the edge of each scanning.

5.6 MEMS micromirror

Rotating is a popular scanning solution for automotive LiDARs and many commercial LiDARs adopt either entirety rotating or mirror rotating as scanners. The rotating scheme provides straight and parallel scanning lines with a uniform scanning speed over a wide FOV. However, rotation-based scanners generally have drawbacks of being generally bulky, with high power consumption and limited frame rate. MEMS-based scanning micromirrors have recently received significant interest in automotive LiDARs due to their merits of being lightweight, compact and with low power consumption. MEMS is short for micro-electro-mechanical systems, which are made up of components between 1 and 100 μm in size. A MEMS LiDAR system uses a tiny mirror whose tilt angle varies when applying a stimulus such as a voltage or a current [5]. The tiny mirror inside is a MEMS scanning micromirror that consists of a silicon device with a millimeter-scale mirror at the center [6]. The mirror is typically connected to flexures that allow it to oscillate on a single axis or biaxially to project or capture light, as illustrated in figure 5.9. A MEMS micromirror that operates on a single axis is far less complex to manufacture and to operate than a biaxial mirror, and it is also far more robust to vibration and shock. Before finding application in LiDAR, biaxial MEMS micromirror has been used in projectors, due to its tiny size, low cost, and fast vibration rate [7].

From its operating mode, a MEMS micromirror can be categorized into resonant and non-resonant. The resonant MEMS micromirror provides a large scan angle at a high frequency and a relatively simple control scheme while the scan trajectory is sine-like, i.e., non-uniform scan speed. A resonant MEMS micromirror resonates at its characteristic oscillation frequency, which is determined by its mass, structure, and spring constant. Using resonance makes it possible to obtain large mirror deflection angle very quickly with a small stimulus. However, a MEMS scanner based on a nonlinear oscillator causes softening and/or stiffening, which limits its operating frequency. A non-resonant MEMS micromirror, that is also referred to as

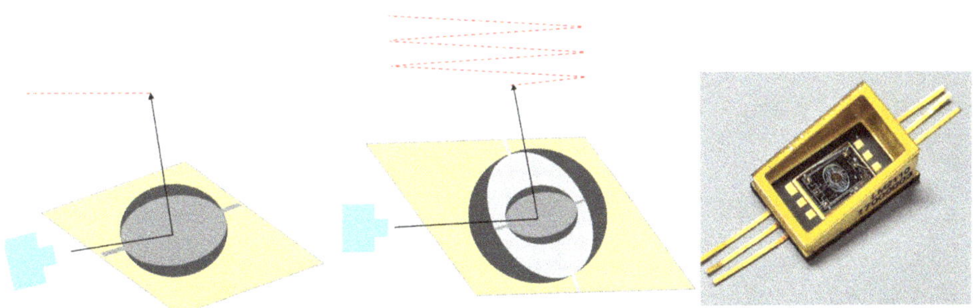

Figure 5.9. Single-axis and biaxial scanning of a MEMS micromirror.

a quasi-static MEMS micromirror, provides a large degree of freedom in the design of trajectory, such as triangular or saw tooth scanning with constant scan speed. The non-resonant mode is used for linear operation that takes advantage of the excellent linearity between the drive current and optical deflection angle. Therefore, the resonant mode is also called nonlinear mode, and the non-resonant mode is also called linear mode. But, to keep the scan quality of non-resonant MEMS micromirror, a rather complex controller is required. Compared to a non-resonant MEMS micromirror, a resonant MEMS micromirror has a wider scanning angle and is energy-efficient. With additional optics and lens, the scan angle of non-resonant MEMS micromirror also can be enlarged. Similar to rotating mirror, the optical deflection angle is twice the mechanical angle, i.e., the tilt angle of mirror.

As for the principle of MEMS micromirror illustrated in figure 5.10, electromagnetic driven for instance, a MEMS mirror consists of a mirror chip and a magnet. The mirror chip includes a mirror, coil and torsion bars. It is formed as a thin film on a portion of a silicon substrate using MEMS technology. An electromagnetic MEMS micromirror is configured with a small but powerful magnet surrounding or beneath the mirror chip to provide an optimal magnetic field to the coil around the mirror while achieving compact size. The surface of the mirror is ultra-smooth to highly reflect the laser beam. The metal coil is deposited on the edge of the mirror to carry current. When current flows through the coil, a torque is produced to tilt the mirror by Lorentz force. The mirror is supported by a pair of torsion bars, which serve as the axis of rotation and torsion bar spring. When the mirror is tilted, an elastic force of the torsion bar spring is applied in the opposite direction to suppress mirror rotation. The change of the mirror's angle is implemented by varying the magnitude of the current flowing though the coil to control the torque. The electromagnetic micromirror offers excellent linearity of the tilt angle versus the applied signal amplitude both in static and dynamic operation. Therefore, the control of an electromagnetic MEMS micromirror is very simple. Electromagnetic actuation is a current driven manner that requires low voltage, below 5 V, making it compatible with standard complementary

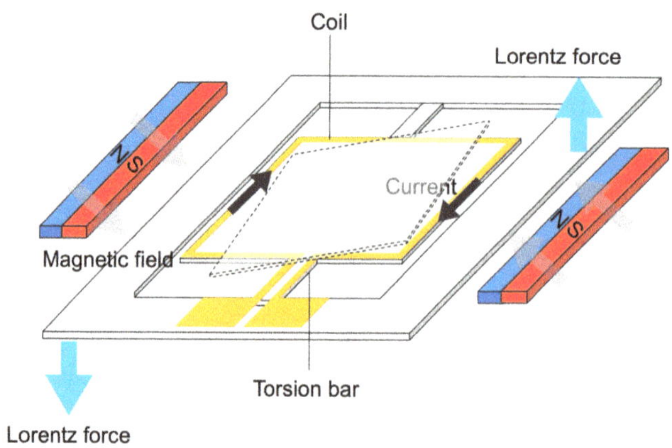

Figure 5.10. Principle of electromagnetic-driven MEMS mirror.

metal-oxide-semiconductor (CMOS) voltage. Also, the electromagnetic MEMS micromirror has durability and low power consumption.

The required drive forces for the mirror movement can be provided by various physical principles. Except for the above-mentioned electromagnetic effect, the relevant principles in practice for driving such a mirror include electrostatic and piezoelectric effects. In contrast to electromagnetic actuation, the resulting drive force of electrostatic actuation between the drive structures cannot be reversed in polarity. For the realization of non-resonant components with positive and negative effective direction, two drives with positive and negative polarity are required. These highly nonlinear drive characteristics in some parts of the deflection area hinder controlling the mirror properly. So, electrostatic micromirrors usually utilize a resonant mode of operation, where an eigenmode is activated. The frequency that relies on the design of MEMS structure, cannot be varied by the driven signal. Piezoelectric actuation produces a high force like an electromagnetic one, but its stroke length is short. To achieve the desired angle, some mechanism utilizing mechanical amplification will be required. This technology is newer than electromagnetic and electrostatic ones and remains to be implemented in commercial products. Specifications of the three driven principles are listed in table 5.1. Electromagnetic actuation supports larger mirror size than the other methods. With similar principle but different fabrication process, scanners with cm-level mirrors can be achieved, making assembly and test easy. The large size of mirror also enables higher efficiency of collection on returned light.

5.7 Galvanometer mirror

A galvanometer mirror is an apparatus like a large-size MEMS micromirror, of which mirror size is cm-level. Different from a MEMS micromirror, a galvanometer mirror can only swing in one dimension. A galvanometer-based scanner uses a pair of cm-level mirrors orthogonal to each other, as shown in figure 5.11. A galvanometer mirror is an ammeter that indicates it has sensed an electric current by deflecting a light beam with a mirror [8]. It consists of a long fine coil of silk-covered copper wire. In the heart of that coil, within a little air-chamber, a small mirror is hung by a single fiber of floss silk, with four tiny magnets cemented to its back.

Table 5.1. Driven principles of MEMS mirror.

Drive method	Electromagnetic	Electrostatic	Piezoelectric
Rotation torque	High	Low	High
Drive voltage	—	50–150 V	20–50 V
Drive current	Approx. 20 mA	—	—
Power consumption	Low	Low	High
Stroke length	Long	Medium	Short
Optical deflection angle (nonlinear mode)	±25°	±15°	±25°
Optical deflection angle (linear mode)	±15°	±5°	Difficult to control
Frequency	Low	Medium	High
Mirror size	Large	Small	Small

When a current traverses the long wire of the coil, the suspended magnets twist themselves horizontally out of their former position, the mirror is inclined with them. Hence, the laser beam deflected on the surface of the mirror scans with the swing of the mirror. When there is no current through the coil, the mirror remains stationary at its zero position. Galvanometer mirrors have merits of fast speed, low step response time, high positional accuracy and high linearity. Close-loop controller feedback of mirror position can be used to optimize galvanometer inputs to improve linearity and minimize positional errors [9]. Besides, close-loop galvanometer mirrors enable a large FOV with high linearity and adjustable speed. Galvanometer mirrors have higher power consumption and larger footprint than MEMS micromirrors, making them LiDAR scanners only in academic papers rather than industrial mass-production LiDARs.

5.8 Risley prism

Risley prism provides another simple yet versatile optical scanning method. A Risley prism scanner utilizes two wedge prisms that are packed sequentially with a coaxial design to steer the light beam [10]. A wedge prism can be considered as an optical window with one side ground and polished at an angle in order to refract, or bend, the input light beam. As with round wedged prisms shown in figure 5.12, the wedged prisms are effectively cylinders with one surface perpendicular to its curved side that is referred to as the flat face, and a surface opposite to the flat face that is referred to as the wedged face. The wedged face is ground with a physical angle. The angle that the beam creates on either side of the interface with respect to the normal of the interface can be determined by Snell's law. The second prism helps extend the scanner's FOV and trajectory's coverage. To rotate the Risley prism, the prisms are mounted inside brushless hollow-motors attached with periodically-poled magnet rings. Current in the external stator coils is supplied and controlled by the respective motor drivers to rotate the prisms at desired rotation speeds and phases. Each prism can rotate about the optical scan axis resulting in the incident beam being refracted by the prisms. The emergent beam is deviated in a direction according to the relative orientation of the prisms with respect to each other. The two prisms are identical in refractive index and wedge angle, meaning the two vectors will have equal

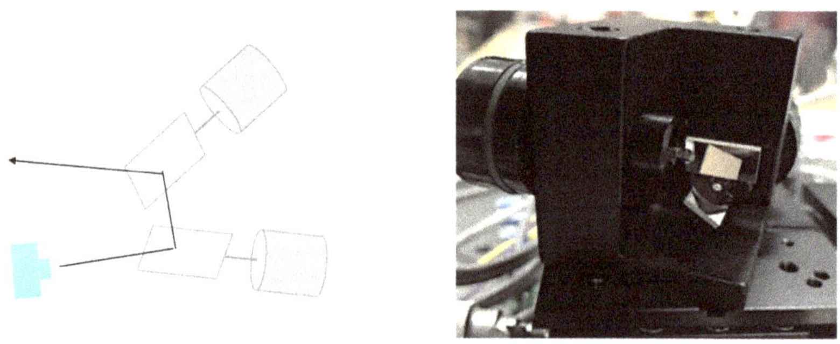

Figure 5.11. Galvanometer mirror.

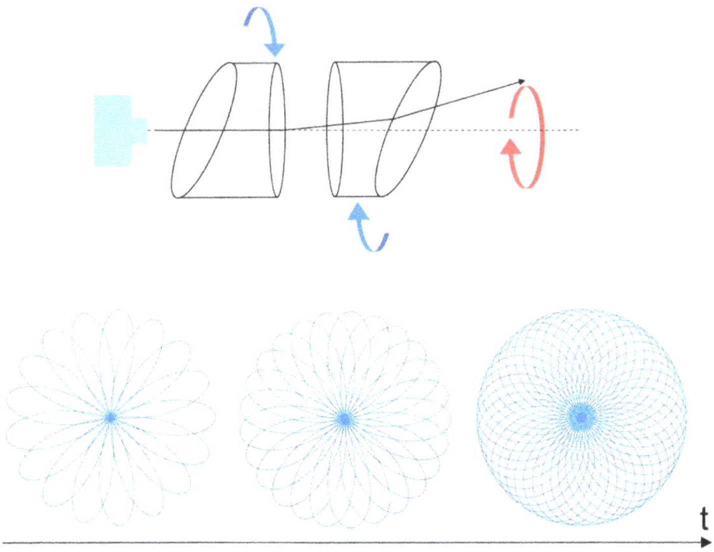

Figure 5.12. Risley prism and its scanning pattern.

magnitude. By rotating the two prisms in different speeds and directions, a large variety of unique patterns can be generated [11]. A rosette pattern scanned by a Risley prism is as shown in the bottom part of figure 5.12. In this pattern, the scanning density distribution becomes retina-like. The pattern is first sparse and then becomes denser than the equivalent 64 beams with long accumulation time. Each scan goes through the center of FOV, leading to repetitive coverage at the center. The returned beam goes back along the same optical path. The orientated view of angle of the scanner for the returned beam is simulaneously the same as the emission, making the efficiency of collection more than MEMS scanning due to its large aperture. Instead of using a single pair of LD and avalanche photodiode (APD), dies of LD array and APD array can be packaged to increase the scanning density. *DJI*, a leading manufacturer of drones based in China, released the first LiDAR on the basis of Risley prism scanner in 2019. The *Livox* LiDAR series released by *DJI* uses a straightforward approach that enables the mass-production LiDAR to be sold at a lower cost, within a range from 599 to 1599 USD (online price for individual customers, price for industrial partners will be lower). As the pattern is quite different from a general raster one, algorithms need additional preprocessiong or optimization to efficiently process the generated point cloud data.

5.9 Wavelength tuning

From this part, we start to go into the solid-state scanners, including the flash approach that illuminates the whole scene without scanning point by point or line by line. Solid-state scanners contain totally no moving parts, making scanners more stable and reliable in harsh environments. In a wavelength tuning approach, the direction change is accomplished by tuning the wavelength of input beam rather

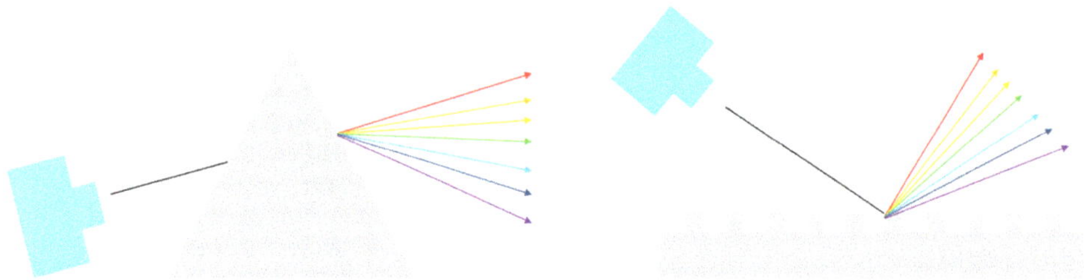

Figure 5.13. Deflection by prism and diffraction grating.

moving any part inside the scanner. As with the prism shown in figure 5.13, in optics, a dispersive prism is an optical prism that is used to disperse light, that is, to separate light into its spectral components, i.e., different colors [12]. Different wavelengths (colors) of light will be deflected by the prism at different angles. Generally, longer wavelength undergoes a smaller deviation than shorter wavelength. The degree of bending of the light's path can be determined by Snell's law. Similar beam steering can be realized by adopting a diffraction grating, which is an optical component with a periodic structure that diffracts light in to several beams travelling in different directions [13]. The directions or diffraction angles of beams depend on the incident angle of the beams to the diffraction grating. In this situation, the grating acts as a dispersive element like the above-mentioned prism to deflect incident light. Therefore, in this way, the beam can be steered to point in any direction by changing its wavelength. Although moving parts are removed, the mechanism to tune the wavelength of laser is required, which increases the complexity or cost of LiDAR. This scanning approach is adopted by *Baraja*, a LiDAR startup based in Australia, which released its scanning technique called Spectrum-Scan.

5.10 Optical phased array

Optical phased arrays (OPAs) represent a technology first demonstrated in the early 90s but becomes attractive in LiDARs with the fast development of autonomous vehicles, which makes simple, affordable and lightweight scanners possible [14]. OPA-based scanners offer very precise stabilization, random-access pointing, programmable multiple simultaneous beams and a dynamic focus/defocus capability. An OPA is the optical analog of a radio-wave phased array, introducing its idea of steering electromagnetic beams and controlling their direction using phase delay and interference between several transmitting antennas into optics [15]. Inference is the basic principle of OPA to implement beam steering, which is a result of coherent superposition of complex-valued wave functions. In a phased array transmitter, interference is induced by electromagnetic waves (light waves) that are emitted by multiple arrayed antenna elements, as illustrated in figure 5.14. The phase of propagating sinusoidal wave changes with propagating distance. When multiple sinusoidal propagating waves with identical frequency reach the same point, they will combine constructively or destructively depending on their relative phases. The constructive (in-phase) and destructive interference (out-of-phase) of waves result in peaks and nulls in the arrayed

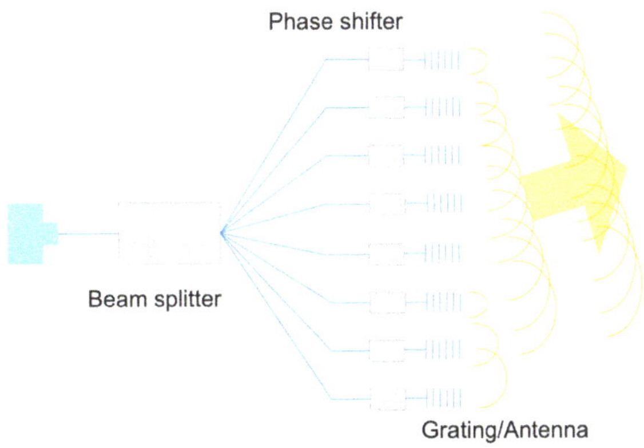

Figure 5.14. Schematic diagram of OPA.

pattern. The peaks in the arrayed pattern can be adjusted to maximize the signal strength transmitted to or received from specific desired locations [16].

The fabrication of OPA can be based on III/V compound semiconductor or silicon photonics. Silicon photonics is more attractive due to its compatibility with standard CMOS process. With a rapid growth of silicon photonics, OPA-based silicon photonics possesses a significant position. Except for laser source, an OPA consists of waveguides, beam splitters, phase shifters and gratings. A waveguide is used to confine the light propagating inside with low loss by the index contrast and connect between different components. It is the most fundamental component in integrated photonics. A beam splitter is a component to split a beam into multiple separate beams, which is usually realized by MMI (multimode interferometer) or star coupler. A phase shifter is a key component in OPA to achieve the redirection of beams by introducing different phase delays to the signals in antennas. In silicon photonics, phase delay is mainly implemented using thermo-optic effect and electro-optic effect to vary the refractive index of silicon waveguides. Therefore, the speed of light propagating inside the waveguide is varied, achieving the phase variation of signal. In the thermo-optic approach, the refractive index of silicon depends on the temperature and can be changed by the variation of the temperature. It is a simple way to implement phase delay that only a heater (electrode) is deposited on the top of the silicon waveguide to heat it by the current flow, while the electro-optic effect varies the refractive index of silicon waveguide by charge injection. The varying speed of the electro-optic effect is faster than the thermo-optic effect, but its fabrication is more complicated due to the dopant of waveguide. The antennas here are similar to those in radio-wave which are used to emit waves, but are in nanoscale size in OPA. They are usually implemented in the form of grating or edge or end-fire couplers, which are placed in an array that is designed delicately by simulation to achieve optimal performance. The design parameters that can be used to optimize interference in order to enhance the OPA performance are in terms of maximum steering along two axes, beam directionality, suppression of side lobes and reduction of crosstalk and emitter power. Although the OPA approach offers advantages of cost-efficiency, tiny footprint, total solid-state and robustness, there have been no successful industrial mass-production LiDAR products based on OPA scanner released. Both academic and industrial researchers are paying much attention to OPA-based LiDARs, but the success of phased array in radio-wave is difficult to be copied by photonics due to issues of loss, nonlinearity and side lobes.

5.11 Flash LiDAR

A flash LiDAR is similar to a camera that receives light from the whole front scene, but adopts an active manner to obtain distance information, as shown in figure 5.15. In flash LiDAR, the entire FOV is illuminated with a wide diverging laser beam in single pulse, which is in contrast to conventional scanning LiDAR. In the above-mentioned scanners, each of them uses a collimated laser beam to illuminate a single point at a time and scan the full view point by point. Whether they are mechanical, hybrid solid-state or entirely solid-state, they need to scan the scene point by point. However, flash LiDARs are fundamentally different from them. Like a camera, a flash LiDAR detects the whole scene frame by frame. Its receiver consists of a large array of APDs, each APD can be considered as one pixel that records distance information individually. When a wide diverging laser beam illuminates the scene, the returned light is simultaneously collected by all the pixels to generate a frame of distance map. Without moving parts, all the components can be integrated or partially integrated using integrated circuits or integrated optoelectronics technology to significantly reduce the size and cost of LiDAR. Its fully solid-state structure also provides reliable and robust operation in harsh environments. In our opinion, this would be the ultimate solution of automotive LiDAR. But flash LiDAR has not been widely used yet due to complicated circuit design and fabrication process. Since each pixel requires one set of receiver and timing modules, integrating a large amount of pixels with these modules while achieving an excellent yield is quite challenging. *Ibeo* and *Continental* have launched solid-state flash LiDAR products for autonomous vehicles. It is not hard to increase the vertical resolution, but a flash LiDAR has either small horizontal FOV or low horizontal angular resolution compared to scanning approaches, due to limited pixels in the horizontal.

Figure 5.15. Flash LiDAR. The ibeoNEXT Generic-Sensor image on the right has been obtained by the author from the Wikimedia website where it was made available by Lidar.HH under a CC BY-SA 4.0 licence. It is included within this article on that basis. It is attributed to Lidar.HH.

References

[1] Raj T, Hashim F H, Huddin A B, Ibrahim M F and Hussain A 2020 A survey on LiDAR scanning mechanisms *Electronics* **9** 741
[2] 2018 LiDAR Sensor Functionality and Variants White paper *SICK AG*
[3] *ALASCA User Manual* 2006 (Hamburg: Ibeo Automobile Sensor GmbH)
[4] McManamon P 2019 *LiDAR Technologies and Systems* 1st edn (Washington, DC: SPIE)
[5] Yoo H W, Druml N, Brunner D, Schwarzl C, Thurner T, Hennecke M and Schitter G 2018 MEMS-based LiDAR for autonomous driving *Elektrotech. Inf.tech.* **135** 408–15
[6] https://en.wikipedia.org/wiki/Microscanner (accessed in January 2022)
[7] Holmström S T S, Baran U and Urey H 2014 MEMS laser scanners: a review *J. Microelectromech. Syst.* **23** 259–75
[8] https://en.wikipedia.org/wiki/Mirror_galvanometer (accessed in January 2022)
[9] Tang E M and Tao Y K 2021 Modelling and optimization of galvanometric point-scanning temporal Dynamics *Biomed. Opt. Exp.* **12** 6701–16
[10] Liu Z, Zhang F and Hong X 2022 Low-cost retina-like robotic LiDARs based on incommensurable scanning *IEEE/ASME Trans. Mechatron.* **27** 58–68
[11] Brazeal R G, Wilkinson B E and Hochmair H H 2021 A rigorous observation model for the Risley prism-based Livox mid-40 LiDAR sensor *Sensors* **21** 4722
[12] https://en.wikipedia.org/wiki/Dispersive_prism (accessed in February 2022)
[13] https://en.wikipedia.org/wiki/Diffraction_grating (accessed in February 2022)
[14] McManamon P F, Dorschner T A and Corkum D L *et al* 2021 Optical phased array technology *Proc. IEEE* **84** 268–98
[15] Hsu C, Li B, Solano-Rivas B and Gohil A R *et al* 2021 A review and perspective on optical phased array for automotive LiDAR *IEEE J. Sel. Top. Quantum Electron* **27** 8300416
[16] Hashemi H 2021 A review of semiconductor-based monolithic optical phased array architectures *IEEE Open J. Solid-State Circuits Soc.* **1** 222–34

IOP Publishing

ToF LiDAR for Autonomous Driving

Wei Wei

Chapter 6

Automotive LiDAR players

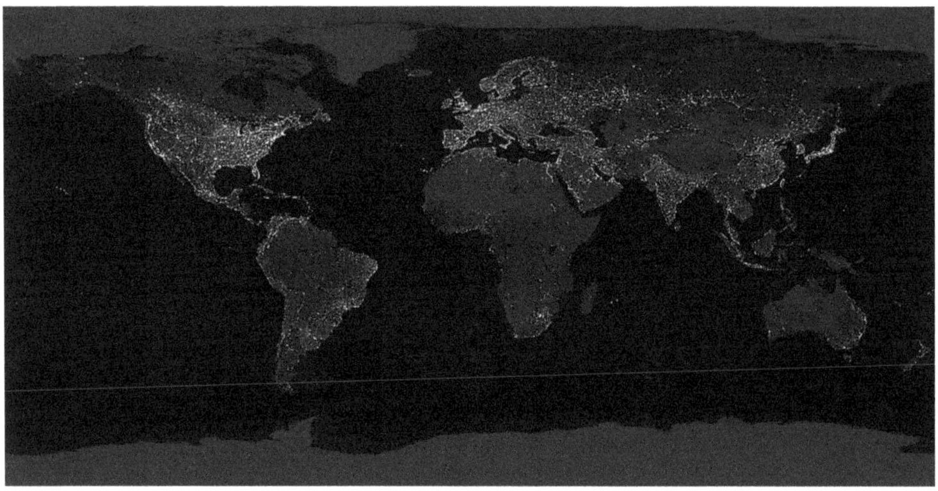

This Earth's City Lights by DMSP image has been obtained by the author from the Wikimedia website https://commons.wikimedia.org/wiki/File:Earth%27s_City_Lights_by_DMSP,_1994-1995_(large).jpg, where it is stated to have been released into the public domain. It is included within this article on that basis.

Ten years ago, automotive LiDAR was a luxury device for participants in autonomous driving, and there were not many choices. Nowadays, plenty of automotive LiDAR players all over the world focus on cutting-edge LiDARs seeing farther, faster, with more accuracy and also with reduced cost and size. There exist many choices for customers whatever kind of LiDAR they need, high-performance, tiny or low-cost. So, in this chapter, global autonomous LiDAR players will be surveyed to help readers get more insight into the LiDAR industry.

doi:10.1088/978-0-7503-3723-6ch6

6.1 Global players

In the past decade, LiDAR became an emerging industrial area that attracted numerous players including startups that focus on LiDAR, sensor manufacturers, Tier-1 suppliers, and even telecommunications companies like *Huawei* and Drone manufacturers like *DJI* (*Livox*). The key players over the world are demonstrated in figure 6.1. The flourishing of the LiDAR industry is accompanying the development of autonomous driving. Most LiDAR players target the huge market of autonomous vehicles. LiDAR companies in different regions have their own characteristics. In USA, influenced by the atmosphere in Silicon Valley, most American LiDAR players are startups founded in the last 10 years and sensitive to new technologies. Unlike USA players, key European players are traditional sensor manufacturers and Tier-1 automotive suppliers. They have mass-producible capabilities, mature marketing and sales teams, and close relationships with automobile manufacturers. So, it is not hard to find customers for European players, but there could be huge challenges for American startups. High-resolution LiDAR for vehicles is oriented from *Velodyne*, a company based in USA. So, let us start to know these key players from the USA region then move to other regions.

6.2 USA

Velodyne is a famous LiDAR technology company, based in San Jose, California, located in the Silicon Valley area. It was founded by David Hall in 1983 as an audio company specializing in subwoofer technology. The company started business on LiDAR from laser distance measurement in 2005 inspired by the DARPA Grand Challenge. In that Challenge, all the robot vehicles on the route had mounted LiDAR sensors to perceive the environment and navigate. The winner

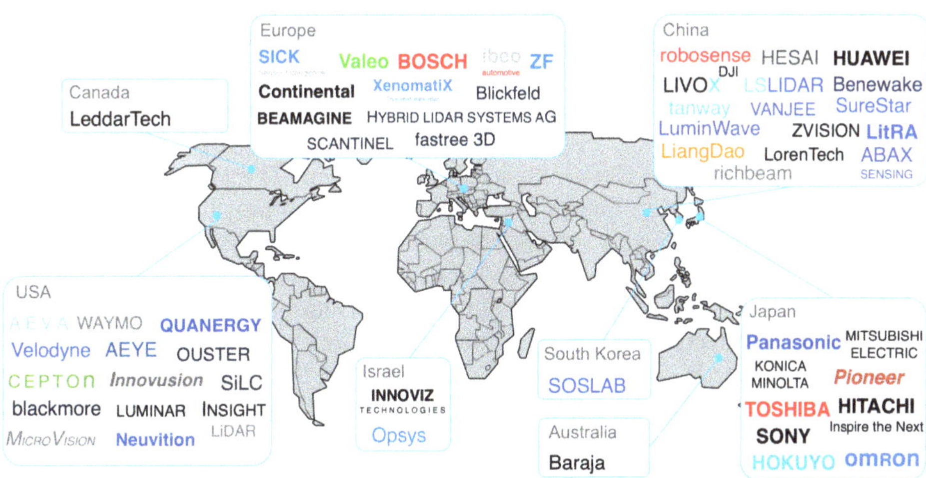

Figure 6.1. Global LiDAR players. Background world map: This Simple world map edit image has been obtained by the author from the Wikimedia website where it was made available by Ashre under a CC BY-SA 4.0 licence. It is included within this article on that basis. It is attributed to Ashre.

Stanley utilized five roof-mounted *SICK* AG 2D LiDAR sensors. David Hall and his brother Bruce noticed the shortcoming of existing LiDAR, which scanned only a single, fixed line of sight. After two years, *Velodyne* developed new LiDAR with 64 lasers rotating to scan the environment for the 2007 DARPA Urban Challenge, and helped five teams finish the race. The new LiDAR created a 360° 3D map of the environment, and produced one million data points per second over the 5000 data points per second of the earlier LiDARs. Since then, *Velodyne* has been famous and popular in autonomous vehicles. In 2010, *Google* began testing autonomous vehicles on the streets in the San Francisco Bay Area using *Velodyne*'s LiDAR. A high-resolution LiDAR from *Velodyne* on the roof of vehicles has almost become a symbol of autonomous driving. Although the price is extremely high, most autonomous driving companies want one mounted on their testing vehicles. In 2016, *Velodyne*'s LiDAR department was spun off from *Velodyne* Acoustics as *Velodyne* LiDAR, and raised \$150M investment from Ford and Baidu. At the beginning stage of autonomous driving, *Velodyne* dominated the LiDAR area, but with more and more players appearing, it did not possess a dominant position in the market anymore. In 2019, *Velodyne* cut its business in China, and closed a direct sales team in the country which was set up in 2015. In November 2022, it was merged with *Ouster*, a young LiDAR technology company based in San Francisco.

Velodyne's main products including HDL-64E, HDL-32E, VLP (Puck), VLS. HDL-64E is the most famous one that started an industry. It was HDL-64E (left in figure 6.2) that was mounted atop five of the six vehicles that finished the race in the 2007 DARPA Urban Challenge and *Google*'s self-driving cars in 2010. It was a symbol of LiDAR and early-stage autonomous vehicles. Over 1200 have been sold, covering 20 countries in North and South America, Europe and Asia. *Velodyne* announced the retirement of HDL-64E in 2019. HDL-64E adopts a typical mechanically rotating mechanism to provide high frame rate ranging from 5 to 20 Hz (depending on different versions) with its full 360° horizontal FOV and 26.8° vertical FOV. Its patented one-piece design uses 64 fixed-mounted lasers to achieve a vertical resolution of 64 lines, each mechanically mounted to a specific vertical angle, with the entire unit rotating. The data is more than 1.3 million points per second, providing dense information. HDL-64E employs the time-of-flight (ToF) principle to measure distance up to 120 m, its accuracy is <2 cm. The laser transmitters are typical pulsed semiconductor laser diodes operating at 905 nm. Then HDL-32E and Puck series were released with smaller volume and lower resolution to provide cost-effective choices. Velarray is the newest LiDAR series released by *Velodyne*. Unlike previous product series, Velarray is solid-state that adopts *Velodyne's* micro-lidar array (MLA) architecture to make its volume very tiny.

Ouster is a LiDAR company founded in 2015, in San Francisco, focusing on high-resolution and digital 3D LiDARs. To be different from most LiDAR companies, *Ouster*'s LiDAR operates at the wavelength region of 850 nm using VCSEL as laser source. The solar spectral photon flux at 850 nm is 2× higher than at 905 nm, up to 10× than at 940 nm, and up to 3× than at 1550 nm, which are typical operating wavelengths of LiDAR systems. But, the wavelength of 850 nm has unique advantages over other wavelengths. Normally, the humid and foggy

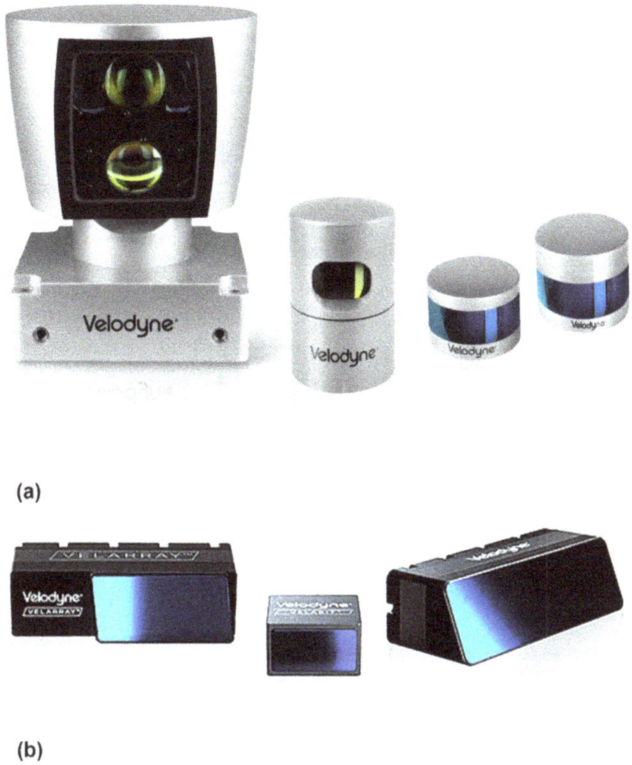

Figure 6.2. *Velodyne* product family. Upper: This Velodyne ProductFamily BlueLens 32GreenLens.png image has been obtained by the author from the Wikimedia website where it was made available by APJarvis under a CC BY-SA 4.0 licence. It is included within this article on that basis. It is attributed to APJarvis. Lower: This Velodyne Lidar Velarray H800 Velabit Velarray M1600 Sensor Family.jpg image has been obtained by the author from the Wikimedia website where it was made available by APJarvis under a CC BY-SA 4.0 licence. It is included within this article on that basis. It is attributed to APJarvis.

conditions that laser pulses transmit through result in less laser light captured by the receiver. 850 nm has much lower atmospheric water vapor absorption than typical wavelengths, especially 1550 nm. Moreover, the sensitivity of silicon detectors at 850 nm is much higher than at long wavelengths, enabling longer measurable distance. *Ouster*' LiDARs use customized arrayed vertical-cavity surface-emitting lasers (VCSELs), single-photon avalanche diode (SPAD) and patented microoptics to realize purely solid-state and improve the performance. Ouster announces on its website that it is the first company to commercialize a high performance SPAD and VCSEL approach.

Quanergy is a Silicon Valley-based technology company, which was founded in 2012, in Sunnyvale, California. To our best acknowledge, *Quanergy* was the first company that announced focusing on OPA technique and releasing OPA-based LiDAR with the price down to $100. At that time, OPA was a very promising and attractive technique to realize solid-state LiDAR with no moving parts. The scanning module could be made at chip scale using complementary metal-oxide-semiconductor (CMOS)-compatible process to shrink the bulky volume of traditional LiDARs and cut down the huge cost. Although some companies followed the same technical route with *Quanergy*, OPA-based LiDAR is very challenging to realize mass production due to high optical loss and nonlinearity. Until 2021, *Quanergy* announced that it delivered industry-first OPA-based solid-state LiDAR using scalable CMOS silicon manufacturing process for cost-effective, mass-market production. But it seems that it missed the golden market opportunity for autonomous driving. Other competitors rapidly dominated LiDAR's market by releasing mass-production products based on mechanical rotating, micro-electro-mechanical (MEMS) mirrors or rotating prisms. In December 2022, *Quanergy* filed for chapter 11 bankrupcy.

Luminar is a LiDAR technology company founded in 2012, with headquarters in Orlando, Florida. *Luminar* first revealed its prototype in April 2017. *Luminar's* Iris LiDAR operates at the wavelength of 1550 nm, so most probably it measures distance in the manner of frequency-modulated continuous wave (FMCW). Fiber laser is adopted as laser source to get a high-energy laser beam at 1550 nm. To receive the returned laser, it uses InGaAs detectors with very high sensitivity and dynamic range rather than conventional silicon detectors. Two-axis scanning mirrors are employed as scanners to obtain a field of view (FOV) of 120° × 28° and a camera-like resolution >300 points per square degree. It also customizes mixed signal ASIC chips to improve performance while significantly reducing cost. The maximum range can be 600 m for larger and brighter objects, and 250 m for small and less reflective objects. *Lunimar* has built partnerships with *Mercedes-Benz* and *Volvo* to accelerate the development of future highly automated driving technologies for partners.

Innovusion is a Silicon Valley-based LiDAR company founded in 2016 that focuses on image-grade LiDAR for autonomous vehicles and advanced driving assistance systems (ADAS). The architecture of *Innovusion*'s LiDAR is similar to *Luminar*'s. A 1550 nm fiber laser is used as emitter and two-axis mirrors are used as a scanner to achieve long measurement range and wide FOV. According to the revealed patent, one-axis steering is realized by a rotating polygon mirror with multi-facet. The other axis steering is accomplished by a scanning optic like MEMS micromirror, galvanometer mirror or polygon mirror. Funded by *NIO* Capital, *Innovusion*'s Falcon LiDAR has been chosen as the standard configuration for the *NIO* ET7 autonomous driving system.

Aeva is a LiDAR company founded in 2017, Silicon Valley, focusing on 4D LiDAR technology for automated driving and industrial applications. *Aeva* LiDAR operates at a wavelength of 1550 nm to measure distance in a manner of FMCW. DFB diodes are employed as laser source. The output laser is coupled into a fiber and then amplified by an EDFA (erbium doped optical fiber amplifier). The scanner also consists of a rotating polygon mirror. But, the target of *Aeva* is shaping the future of sensing and perception with its 4D LiDAR-on-chip technology that incorporates all key sensor components onto a silicon photonics module, with no fiber optics. *Aeva* became a publicly listed company in March 2021 after completing a business combination with *InterPrivate Acquisition*, a publicly traded special purpose acquisition company (SPAC).

Aurora (***Blackmore***) is self-driving vehicle technology company based in Pittsburgh, Pennsylvania and Mountain View, California. *Blockmore* is a sensors technology company that pioneers behind Doppler LiDAR (FMCW) for automotive applications. *Blockmore* announced the world's first Doppler LiDAR for autonomous fleets with instantaneous velocity and range data for objects beyond 450 m in January 2019. In May 2019, *Blockmore* was acquired by *Aurora*, the latter integrated *Blackmore*'s LiDAR into its self-driving stack.

Cepton is a Silicon Valley-based technology company founded in 2016 with a focus on LiDARs for ADAS in mass-market consumer vehicles. It sought to create high-performance, reliable, and low-cost LiDAR sensors using its patented Micro Motion Technology (MMT) technology and 905 nm laser source. MMT is a mirrorless, rotation-free and frictionless method for 3D LiDAR imaging. MMT-based LiDARs are not entirely solid-state, their scanner contains two sets of platforms driven in a mechanical manner. Its products are featured in high resolution up to 0.05° × 0.05° (ROI), long distance up to 200 m at 10% reflectivity and wide FOV of 120° × 25°. *Cepton* developed an early relationship with Japanese lighting supplier *Koito* and won a contract to supply *General Motors* with its sensors for various future production vehicle programs in 2021. It became a publicly listed company in February 2022 after completing a business combination with *Growth Capital Acquisition Corp.*, a publicly traded special purpose acquisition company.

AEye, founded in 2013, Dublin, California, develops AI-powered LiDAR systems for autonomous vehicles, ADAS, and robotic vision applications. It created a long-range LiDAR system that combines an amplifiable 1550 nm laser with a MEMS-based scanner. An amplifiable 1550 nm laser is most probably a fiber laser amplified by a compact EDFA. According to *AEye*'s claim, its LiDAR can detect vehicles at a distance of 1000 m and people at a distance of up to 200 m. *Continental*, an international Tier-1 supplier in Germany, integrated AEye's long-range LiDAR technology into its full stack automated and autonomous solution.

Waymo is a significant player that differs from the above players as it is an autonomous driving technology company who developed its own LiDAR, dubbed the 'Laser Bear Honeycomb.' *Waymo* was formerly known as the *Google* Self-Driving Car Project and then independently founded as an autonomous driving technology company in 2016. *Waymo* first began manufacturing its own LiDAR sensors in early 2017 as a way to dramatically reduce the cost of its autonomous vehicle business. In 2019, *Waymo* announced that it would begin selling its LiDAR, but it stopped this business in 2021. There is not really much public technical material related to *Waymo*'s LiDAR.

MicroVision is an automotive LiDAR technology company headquartered in Redmond, Washington. Its LiDAR scanner adopts biaxial MEMS based on electromagnetic and PZT actuation technologies. It has been developing MEMS technology since 1997 to meet the cost and size requirements of laser beam scanning applications. Its LiDAR operates at 905 nm to measure distance in a direct ToF manner. The company's integrated approach uses its proprietary technology to provide automotive LiDAR sensors and solutions for ADAS and for non-automotive applications including industrial, smart infrastructure and robotics. In early 2023, *MicroVision* completed the acquisition to *Ibeo*, a German LiDAR technology company, expecting to expand partnership with *ZF*, a famous German car parts maker.

Neuvition, founded in 2016, in Santa Clara, aims at manufacturing and design of LiDAR sensors, and providing customized solutions in autonomous driving, intelligent transportation, 3D mapping, volume measurement, and industrial machine vision. Its global headquarters has now located in Suzhou, China. Its LiDAR is MEMS-based, and operates at 1550 nm, using a fiber laser as source. Until now, it has cases in applications of mining and intelligent transportation, but no autonomous vehicles yet.

Insight LiDAR, founded in 2016, Lafayette, focusing on chip-scale, ultra-high resolution FMCW LiDAR sensors. According to its patent disclosure, Insight employs a diffractive element and a widely tunable laser to achieve beam steering. It puts all laser transmission, control and detection on optical semiconductors or photonic integrated circuits (PICs) to reduce the loss and usage of fiber.

SiLC, founded in 2018, in Monrovia, is a technology company that provides chip-scale coherent LiDAR solutions based on silicon photonics technology. Its LiDAR operates at 1550 nm and ranges in a FMCW manner. Different from most FMCW source solutions that amplify seed laser using EDFA, *SiLC* adopts semiconductor optical amplifier (SOA) to amplify the seed laser. The gain of a SOA is generally lower than that of an EDFA, but it can be integrated with a PIC based transceiver to achieve chip-scale transceiver. The fiber-pigtailed module is connected to a scanning unit located at different positions.

6.3 Canada

LeddarTech is a Canadian automotive solutions company founded in 2007 as a successful spin-off Canadian leading optics and photonics research institute. At the early stage of *LeddarTech*, it focused on LiDAR related technology and targeted solid-state LiDAR modules. Now, it provides solutions of sensor fusion and perception platform for ADAS and high automated driving. The resolutions of its products are not as high as those of other competitors. The high-end modules Leddar M16 families have only 16 segments. The Leddar uses diffused light from multiple LEDs or LDs to illuminate a scene in short pulses. Sixteen detectors are placed in an array to receive the returned light and segment the scene. The LD version operating at 905 nm is capable of detecting a longer distance than the LED version operating at 940 nm.

6.4 Europe

SICK is a global manufacturer of sensors and sensor solutions for industrial applications, based in Waldkirch, Germany. Unlike the above LiDAR high-tech startups, *SICK* is a traditional global manufacturer of sensors founded in 1946. It is particularly well known for its laser scanners, which were used for range detection of autonomous vehicles in both the 2004 and 2005 DARPA Grand Challenges. The LMS 291 LiDAR gained fame in the 2005 DARPA Grand Challenge. It was the outstanding performance of LMS 291 in the 2005 DARPA Grand Challenge that inspired the founder of *Velodyne* to develop HDL-64E. LMS 291 can measure range out to 80 m over a 180° arc. The beam scanning is accomplished by a rotating mirror 45° tilted. It is a direct ToF LiDAR working at 905 nm. The arrayed LMS 291 LiDARs with different pointing directions helped *Stanley* (figure 6.3), created by the Stanford University Racing Team, to win the race and the 2 million US dollar prize. *SICK* was famous for its 2D LiDARs with only single-line resolution. Now, it has products with multi-line resolution. Its products are widely used in factory automation, logistic automation and process automation.

Valeo is a French global automotive Tier-1 supplier headquartered in Paris, supplying a wide range of products to automakers and the aftermarket. Scala 3D laser scanner (Gen 1) is *Valeo*'s first LiDAR product, also the world's first automotive-grade LiDAR. It operates at 905 nm and contains three avalanche photodiode (APD) elements with an active area of 160 μm × 595 μm. A rotating mirror is adopted as a scanner to obtain a vertical resolution of three layers per scan and four effective layers per two scans. The scanning rate is 25 Hz. The typical detection distance is 150 m. Scala is the first LiDAR that is used by an automaker. At the end of 2017, *Audi* launched its redesigned A8 sedan, of which the key sensor is a Scala LiDAR supplied by *Valeo*. *Audi* announced that it was the auto industry's first Level 3 autonomous car. So, it is a sign that automakers are starting to use LiDARs in/on vehicles.

Bosch is another global automotive Tier-1 supplier headquartered in Gerlingen, Germany. According to *Bosch* Media, *Bosch* is now working on making LiDAR sensors production-ready. Its LiDAR sensor is a long-range one. But there are no detailed specifications.

Figure 6.3. LMS 291 LiDARs on the roof of *Stanley*. This adapted Stanley2.jpg image has been obtained by the author from the Wikimedia website https://commons.wikimedia.org/wiki/File:Stanley2.JPG, where it is stated to have been released into the public domain. It is included within this article on that basis.

Continental is also a German automotive Tier-1 supplier headquartered in Hanover. It is structured into six divisions: Chassis and Safety, Powertrain, Interior, Tires, ContiTech, ADAS. For higher level autonomous systems, *Continental* developed short-, long-range and high-resolution LiDARs named SRL121, HRL131, and HFL110. HFL110 is a solid-state flash unit for Level 2+ and 3 autonomous vehicles. It operates at 1064 nm, which is an unusual wavelength compared to the popularly used 905 nm, 940 nm, and 1550 nm. As a flash LiDAR, it provides a high resolution of 128 × 32 pixels. The short-range LiDAR SRL121 is a point-to-point range finder for emergency brake assist. It operates at 905 nm with a typical pulse peak power of 70 W. HRL131 is a MEMS- based LiDAR operating at 1550 nm for Level 3 and 4 autonomous vehicles. The MEMS scanner is provided by *AEye*'s patented MEMS technique. The detection range is up to 1000 m@high reflectivity, 300 m@10% reflectivity. The mass production of long-range LiDAR will be ready in 2024. *Continental* have built relationships with *Toyota* and *Lexus* to supply its LiDARs for Toyota Mirai 2 and New Lexus LS.

ZF and ***Ibeo*** built close cooperation in LiDAR products. *ZF* is a German famous car parts maker headquartered in Friedrichshafen. *Ibeo* is a German LiDAR technology company founded in 1998. *ZF* holds a 40% stake in *Ibeo*. *Ibeo* has rotating mirror-based 2D LiDAR ALASCA. Its 3D LiDAR ibeoNEXT is a solid-state one with no moving parts. It adopts photo laser measurement technology to detect distance. It is a flash LiDAR with a resolution of 128 × 80 pixels. It consists of VCSELs and SPAD detectors operating at 885 nm, whose wavelength is similar to *Ouster*. *ZF* was granted by *Ibeo* to produce the ibeoNEXT solid-state LiDAR. In Jan. 2021, *Ibeo* and *SICK* announced a new partnership to co-develop their iboeNEXT 3D solid-state LiDAR sensor to make it available for industrial applications. But, *Ibeo* filed for insolvency in September 2022.

Xenomatix, founded in 2013, Leuven, Belgium, develops automotive vision solutions based on solid-state LiDAR technology. It LiDAR is entirely solid-state and employs arrayed VCSELs operating at 840 nm and SPAD. It has compact size, but not as excellent performance as those of competitors in detection range, FOV and resolution.

Beamagine, headquartered in Barcelona, Spain, was created with the goal of commercializing the LiDAR imaging technology developed at CD6. Its scanner adopts MEMS mirror as the manner of beam steering. The ranging principle is direct ToF, but its transmitter operates at a different wavelength of 1064 nm, of which laser source may be a solid-state laser with larger size than a diode laser. The advantage of the wavelength for automotive LiDAR is not very clear for now.

Hybrid Lidar Systems, founded in 2018, Niedersachsen, Germany, develops LiDAR technology for autonomous vehicles and industrial use cases. Its automotive LiDAR LiSSA is solid-state flash LiDAR, operating at 905 nm, and can be customized to work at 850 and 940 nm. The long-range version can detect up to 220 m (10% reflectivity).

Blickfeld, founded in 2017, Munich, Germany, provides LiDAR technology for autonomous mobility and IoT applications. Its scanning adopts a MEMS-based approach. The transmitter operates at 905 nm. Its receiver employs SiPMs, arrays of SPADs connected in parallel, to achieve measurements at lower operating voltages.

Scantinel, founded in 2019, Ulm, Germany, develops FMCW LiDAR technology for autonomous vehicles. It was the first LiDAR company to develop FMCW LiDAR in Europe. The amplification of seed laser is realized by SOA to achieve chip-level integration using PIC. Partly similar to *Quanergy*, *Scantinel* also adopts OPA to steer the direction of beams. But it adopts a hybrid structure containing 1D OPA for one dimension scanning and a rotating mirror for the other dimension scanning, called optical enhanced array (OEA).

Fastree 3D, founded in 2014, Lausanne, Switzerland, is a fabless semiconductor company that focuses on solid-state LiDAR, a spin-off from EPFL. It developed a receiver chip fabricated in a standard CMOS process, which integrates SPADs, time-to-digital converters (TDCs) and digital pre-processing circuitry in one chip. The transmitter adopts arrayed VCSELs as laser source, and is presumed to operate at 905 nm or 850 nm.

6.5 Israel

Innoviz is an Israeli LiDAR technology company founded in 2016. InnovizOne is a solid-state LiDAR specifically designed for automakers and robotaxi, shuttle and delivery companies requiring an automotive-grade, mass-producible solution. It is seamlessly integrable into Level 3–5 autonomous vehicles. It is MEMS-based operating at 905 nm. The detection range is up to 250 m. InnovizTwo is an upgraded version with higher resolution, longer detection and wider FOV compared to InnovizOne. *Innoviz* has built a partnership with *Cariad SE*, *VW*'s automotive software company, to integrate its technology into upcoming *VW* vehicles. In April 2022, *BMW* revealed that *Innoviz*'s LiDAR would be on the 2023 BMW i7 electric vehicles.

Opsys, founded in 2016, Holon, Israel, develops solid-state LiDAR sensors for autonomous applications. It adopts a patented technology called scanning micro-flash. It uses high-density detector arrays coupled with high-density, fully addressable 2D VCSEL arrays to measure distance in a ToF approach without any moving parts. It created a collaboration with Huayu Automotive Systems to supply its solid-state LiDAR.

6.6 China

Hesai is a global LiDAR R&D and manufacturing company, focusing on LiDARs for high-level ADAS and autonomous vehicles. It was founded in 2013 and is based in Shanghai, China. AT128 is a rotating mirror-based high-resolution LiDAR. Its transmitter contains 128 arrayed VCSELs operating at 905 nm. PandarGT is a MEMS-based LiDAR that adopts fiber laser operating at 1550 nm. Pandar128 is a mechanical high-resolution LiDAR. FT120 is a fully solid-state LiDAR with compact size. So, *Hesai* focuses on many popular LiDAR technical routes, including mechanical, MEMS, rotating mirror, 905 nm arrayed VCSELs, and 1550 nm fiber laser. Through the cooperation with Chinese automobile manufacturers of *Li Auto* and *Geely*, *Hesai* got the ticket for mass-production vehicles with autonomous driving systems. In February 2023, *Hesai* became a publicly listed company and began to be traded on Nasdaq.

RoboSense is a provider of smart LiDAR sensor systems for autonomous vehicles, founded in 2014, Shenzhen, China. Similar to *Hesai*, *RoboSense* has rich LiDAR product lines covering mechanical, solid-state and MEMS-based. Its M-series is a featured automotive-grade product for autonomous vehicles. M-series adopts a chip-level 2D MEMS module to realize wide 2D scanning. It operates at 905 nm and detects up to 200 nm. The FOV is 120° × 25°, similar to products from other companies. The average horizontal and vertical resolutions are 0.2°. *RoboSense*'s E-series is automotive-grade solid-state LiDAR that can be used as auxiliary LiDAR for M-series to eliminate blind areas. E-series employs 2D addressable VCSELs as laser source. Recently, *RoboSense* has been officially integrated into *Toyota*'s supply chain system. Prior to this, *RoboSense* had reached partnership with Chinese automakers of *XPeng*, *GAC*, and *Geely*.

Huawei is a global technology company headquartered in Shenzhen, China, focusing on telecommunications equipment and consumer electronics. *Huawei* has been recognized in the automotive industry for its smart car solution and autonomous technology. It provides autonomous driving and smart solutions including LiDARs for Chinese automakers of *AITO*, *BAIC* and *Changan*. There is not much public information about the technical detail of *Huawei*'s LiDAR. Due to its strong capability in manufacturing, *Huawei* most probably has its own LiDAR factory to produce LiDARs.

Livox is an independent LiDAR technology company founded in 2016 through *DJI*'s Open Innovation Program. *DJI* is a world-leading manufacturer of drones headquartered in Shenzhen, China. *Livox*'s first launched Mid-40 and Mid-100 units got the most attention, since they cost only $599 and $1499, respectively, available for everyone from the *DJI* store. It is probably the first 3D high-resolution LiDAR that falls below the price of $1000. *DJI*'s strong capabilities of supply chain management and manufacturing may contribute greatly to the reduction of *Livox*'s LiDAR cost. *Livox* adopts 905 nm ToF technique and a unique scanner scheme of rotating Risley prisms. The transmitter source is six edge emitting lasers (EELs) assembled in parallel. The scanning method features non-repetitive scanning patterns differing significantly form repetitive linear scanning of traditional LiDARs. With longer integration time, the pattern becomes denser, especially the central region. *Livox* categorized Mid series as industrial-grade LiDARs and launched HAP, its first automotive-grade LiDAR. *Livox* is now supplying LiDAR units for *XPENG*, a Chinese automaker in electric vehicles.

LSLiDAR, founded in 2015, Shenzhen, China, provides LiDAR and complete solutions for autonomous driving, intelligent transportation, and smart logistics. It has diverse LiDAR product types, including mechanical, MEMS-based hybrid solid-state, flash, and both ToF and FMCW. LSLiDAR's in-house 1550 nm fiber laser is designed with a master oscillator plus power amplifier structure.

Benewake is a LiDAR company based in Beijing, providing rangefinders and 3D LiDARs. Its short-range LiDAR is a flash one using arrayed VCSELs working at perhaps 850 nm. The long-range LiDARs work at 1550 nm using fiber laser and MEMS mirror for Horn-X1 and rotating mirror for Horn-X2. ***SureStar*** headquartered in Beijing, China, provides LiDARs for navigation and surveying. It adopts a mechanical and micro-rotating mirror to achieve 2D scanning. The products cover both 905 nm and 1550 nm. ***Richbeam***, also headquartered in Beijing, provides industrial single-line and multi-line LiDAR, and automotive-grade LiDAR. It adopts a mechanical and rotating mirror to scan in 2D. The LiDARs operate at 905 nm using VCSEL as laser source. ***Abax sensing*** is a LiDAR chip and sensor company based in Ningbo, China. It provides iToF and dToF chips and VCSEL chips for LiDARs, and solid-state LiDARs. The VCSEL chips operate at 940 nm. ***Tanway***, based in Beijing, provides MEMS-based hybrid solid-state LiDAR operating at 905 nm. The diode lasers are focused in linear shape and then received by multiple elements of APDs. ***VanJee*** based in Beijing, China, provides mechanical and MEMS-based multi-line LiDARs operating at 905 nm. ***LiangDao***, based in Beijing, China, provides solid-state flash LiDAR for the detection of blind areas surrounding vehicles. ***ZVISION***, based in Beijing, China, provides MEMS-based hybrid solid-state LiDAR operating at 1550 nm, using fiber laser as source. ***LuminWave***, based in Hangzhou, China, provides OPA-based LiDAR operating at 1550 nm. ***LitRA***, based in Shenzhen, China, provides mechanical and OPA-based solid-state LiDAR operating at 905 nm. ***LorenTech***, based in Beijing, China, provides solid-state flash LiDAR operating at 905 nm.

6.7 South Korea

SOSLAB was founded in 2016, Gwangju, South Korea, and develops LiDAR sensors for autonomous vehicles, robotics, as well as smart and safer cities with its 2D, 3D LiDAR sensors. Its products cover mechanical 2D LiDAR, MEMS-based 3D LiDAR, and flash 3D LiDAR. The ranging principle is direct ToF that works at the wavelength of 905 nm. Its solid-state LiDAR adopts addressable VCSEL array and 2D SPAD array. In 2022, *Hyundai Motor Group* and *SOSLAB* entered into a memorandum of understanding for the joint development of LiDAR for mobile robots.

6.8 Japan

The LiDAR market in Japan seems not very active compared to other regions like USA and China. It has attracted only traditional companies with versatile business. LiDAR is only a small fraction of their business. However, Japanese companies may possess a very significant position in the global supply chain of LiDAR, due to their large advantages in semiconductors, electronics and mechanics.

Pioneer Smart Sensing Innovations, a consolidated subsidiary of *Pioneer* Corporation, has started mass production of the short-range type of the 3D

LiDAR since November 2020. It is a MEMS-based LiDAR operating at 905 nm. But *Pioneer* withdrew from 3D LiDAR development in early 2022. ***Mitsubishi Electric*** also developed a MEMS-based LiDAR with compact size for safe and secure autonomous driving. ***Panasonic*** developed 3D LiDAR enabling 3D detection of distances with wide FOV for the navigation of autonomous robots. It is a short-range one below 50 nm, which is suitable for autonomous vehicles, but can be used to eliminate blind areas. It is a mechanical LiDAR driven by motors in two perpendicular axes. It also developed an on-chip beam steering device utilizing silicon photonics and liquid crystal technologies. The ranging principle may be ToF, since *Panasonic* has technical advantages in arrayed APD chips. ***Omron*** developed entirely solid-state ToF sensor (ToF camera) for indoor applications and 3D long-range LiDAR for ADAS and autonomous driving. It also has a 2D mechanical laser scanner. ***Konica Minolta*** built a strategic alliance on the development of 3D LiDAR in 2017. ***Hokuyo*** is a global manufacturer of sensor and automation technology, especially LiDAR sensors. It has versatile low-resolution LiDAR types including point-to-point and single-line for autonomous robotics and industrial automation. It adopts 880 nm LED and 905 nm diode laser to measure distance in ToF approach. ***Toshiba*** launched the world's smallest 3D LiDAR for autonomous driving in 2022. It realizes 2D scanning by MEMS and APD array. Although it is palm-size, the detection range can be up to 300 m. However, the FOV is only 24° × 12°, not as wide as MEMS-based or rotation-based ones. By combining more projectors, it can cover wider FOV up to 72°. ***Sony*** has no LiDAR products, but developed SPAD sensor chips, a key component in LiDAR. Its SPAD sensor has supplied for consumer-grade LiDAR in the iPhone. *Sony* also developed SPAD sensor chip for automotive applications, allowing highly accurate and rapid measurement of a distance up to 300 m. In future solid-state LiDARs, *Sony*'s SPAD sensor chip may possess a significant position similar to its image sensor.

6.9 Australia

Baraja is a LiDAR technology company founded in 2015, in Australia. It adopts FMCW technique at 1550 nm to measure distance. Different from all the other LiDAR companies, *Baraja* employs its own Spectrum-Scan solid-state scanning mechanism to steer the direction of a laser beam. According to physical fundamentals, light with different colors goes different angles at the interface between a prism and air. So, this technique requires the wavelength of laser source to tune in a wide range. *Baraja*'s Spectrum HD25 can detect distance up to 250 m. Its FOV is typical 120° × 25°. However, according to public news, *Baraja* has not built solid cooperation with any automaker.

6.10 List of LiDAR for mass production

See table 6.1 for cooperation between LiDAR companies and automakers.

Table 6.1. Cooperation between LiDAR companies and automakers.

Brand	Model	Quantity	Supplier
Audi	A8	1	Valeo
	e-tron	1	Aeva
Mercedes-Benz	S-Class	1	Valeo
BMW	iX		Innoviz
Volkswagen	ID BUZZ	1	Aeva
Honda	Legend Hybrid EX	5	Valeo
Toyota	Mirai	/	Continental
Lexus	LS	/	Continental
Volvo	XC90	/	Luminar
Polestar	Polestar 3	1	Luminar
Lucid Motors	Lucid Air	1	RoboSense
Lotus	Electre	2	RoboSense
BAIC	ARCFOX AlphaS	3	Huawei
XPENG	P5	2	Livox
	P7i	2	RoboSense
	G9	2	RoboSense
NIO	ET7	1	Innovusion
	ET5	1	Innovusion
	ES7	1	Innovusion
Li Auto	L9	1	Hesai
	X01	1	Hesai
GWM	Mocha	3	Ibeo
	Jijialong	4	Huawei
	Blue Mountain	2	RoboSense
SAIC	R-ES33	1	Luminar
	R7	1	Luminar
	L7	2	RoboSense
	LS7	2	RoboSense
Changan	Avatr E11	3	Huawei
GAC	AION LX Plus	3	RoboSense
GM	Ultra Cruise	—	Cepton
Huayu Auto	—	—	Opsys
Geely	Lotus Electre	4	Hesai
			RoboSense
BYD	U8	3	RoboSense
	N7	2	RoboSense
GAC	AION LX Plus	3	RoboSense
	AION Hyper GT	3	RoboSense

Further reading

[1] https://en.wikipedia.org/wiki/Velodyne_Lidar (accessed in January 2022)
[2] https://caixinglobal.com/2019-12-11/velodyne-lidar-cuts-direct-sales-team-in-china-101492860.html (accessed in January 2022)
[3] https://velodynelidar.com/products/velarray-m1600/ (accessed in January 2022)
[4] https://hypertech.co.il/wp-content/uploads/2015/12/HDL-64E-Data-Sheet.pdf (accessed in January 2022)
[5] https://velodynelidar.com/blog/hdl-64e-lidar-sensor-retires/ (accessed in January 2022)
[6] https://en.wikipedia.org/wiki/Ouster_(company) (accessed in January 2022)
[7] https://ouster.com/zh-cn/blog/how-multi-beam-flash-lidar-works/ (accessed in January 2022)
[8] https://quanergy.com/pressrel/quanergy-delivers-industry-first-250-meter-range-for-opa-based-solid-state-lidar/ (accessed in March 2022)
[9] https://spectrum.ieee.org/quanergy-solid-state-lidar (accessed in March 2022)
[10] https://en.wikipedia.org/wiki/Quanergy (accessed in March 2022)
[11] https://en.wikipedia.org/wiki/Luminar_Technologies (accessed in March 2022)
[12] https://sae.org/news/2022/12/luminar-lidar-for-volvo-ex90 (accessed in March 2022)
[13] https://luminartech.com/technology (accessed in March 2022)
[14] https://autonomoustuff.com/products/luminar-iris (accessed in March 2022)
[15] Innovusion Ireland Limited 2020 LiDAR systems that use a multi-facet mirror *US Patent* Application US20200150247A1
[16] https://innovusion.info/products/falcon/ (accessed in March 2022)
[17] https://golden.com/wiki/Innovusion-AMMJ4DX (accessed in March 2022)
[18] https://cnevpost.com/2022/01/06/nio-supplier-innovusion-shows-off-lidar-for-et7-unveils-new-offering-at-ces/ (accessed in March 2022)
[19] https://golden.com/wiki/Aeva%2C_Inc.-PB63KPW (accessed in March 2022)
[20] https://aeva.com/aeries-i/ (accessed in March 2022)
[21] https://businesswire.com/news/home/20210312005301/en/Aeva-and-InterPrivate-Acquisition-Corp.-Announce-Closing-of-Business-Combination-Aeva-to-Trade-on-NYSE-as-AEVA-Beginning-On-March-15 (accessed in April 2022)
[22] https://en.wikipedia.org/wiki/Aurora_Innovation (accessed in April 2022)
[23] https://blog.aurora.tech/progress/aurora-to-acquire-blackmore-industry-leading-lidar (accessed in April 2022)
[24] https://blog.aurora.tech/engineering/fmcw-lidar-the-self-driving-game-changer (accessed in April 2022)
[25] https://businesswire.com/news/home/20190108005293/en/Blackmore-Announces-the-World%E2%80%99s-First-Doppler-Lidar-for-Autonomous-Fleets (accessed in April 2022)
[26] https://techcrunch.com/2019/05/23/fresh-off-a-530m-round-aurora-acquires-lidar-startup-blackmore/ (accessed in April 2022)
[27] https://cepton.com/ (accessed in April 2022)
[28] https://lidarmag.com/wp-content/uploads/PDF/LIDARMagazine_Walker-CeptonTechnolgies_Vol8No2.pdf (accessed in April 2022)
[29] https://investorbrandnetwork.com/clients/cepton-Inc./ (accessed in May 2022)
[30] https://forgeglobal.com/cepton-technologies_ipo/ (accessed in May 2022)
[31] https://eetimes.eu/aeye-lidar-technology-for-autonomous-vehicles/ (accessed in May 2022)
[32] https://en.wikipedia.org/wiki/Waymo (accessed in May 2022)

[33] https://theverge.com/2021/8/27/22644370/waymo-lidar-stop-selling-autonomous-cars-laser-bear (accessed in May 2022)
[34] https://forbes.com/sites/bradtempleton/2021/10/11/new-lidars-from-waymo-and-others-produce-amazing-results/?sh=6ed958b74130 (accessed in May 2022)
[35] https://leddartech.com/about-leddartech/ (accessed in May 2022)
[36] https://vision-systems.com/cameras-accessories/article/16745850/ledbased-leddar-sensor-technology-works-like-lidar (accessed in May 2022)
[37] https://mapix.com/lidar-scanner-sensors/leddartech-lidar/leddar-m16/ (accessed in May 2022)
[38] https://leddartech.com/leddartech-announces-140-million-usd-in-series-d-financing-combined-with-debt-facility/ (accessed in May 2022)
[39] https://en.wikipedia.org/wiki/Sick_AG (accessed in May 2022)
[40] https://daslhub.org/unlv/wiki/lib/exe/fetch.php?media=dylanw:sicklms200.pdf (accessed in May 2022)
[41] https://en.wikipedia.org/wiki/Valeo (accessed in May 2022)
[42] https://hexagondownloads.blob.core.windows.net/public/AutonomouStuff/wp-content/uploads/2020/10/valeo-scala-datasheet-whitelabel.pdf (accessed in May 2022)
[43] https://eetasia.com/teardown-lessons-learned-from-audi-a8/ (accessed in May 2022)
[44] https://en.wikipedia.org/wiki/Robert_Bosch_GmbH (accessed in May 2022)
[45] https://bosch-presse.de/pressportal/de/en/lidar-sensors-the-third-type-of-sensor-technology-needed-for-automated-driving-alongside-radar-and-camera-249985.html (accessed in May 2022)
[46] https://en.wikipedia.org/wiki/Continental_AG (accessed in May 2022)
[47] https://wardsauto.com/vehicles/continental-says-long-range-lidar-key-av-evolution (accessed in May 2022)
[48] https://continental-automotive.com/en-gl/Passenger-Cars/Autonomous-Mobility/Enablers/Lidars/3D-Flash-Lidar (accessed in May 2022)
[49] https://continental-automotive.com/en-gl/Passenger-Cars/Autonomous-Mobility/Enablers/Lidars/Short-Range-Lidar (accessed in May 2022)
[50] https://continental-automotive.com/en-gl/Passenger-Cars/Autonomous-Mobility/Enablers/Lidars/HRL131 (accessed in May 2022)
[51] https://en.wikipedia.org/wiki/ZF_Friedrichshafen (accessed in May 2022)
[52] https://mobilerobotguide.com/2021/01/19/sick-and-ibeo-partner-to-build-new-3d-lidar-solution/ (accessed in May 2022)
[53] https://press.zf.com/press/en/media/media_18881.html (accessed in May 2022)
[54] https://iot-automotive.news/zf-starts-production-of-lidar-sensors/ (accessed in May 2022)
[55] https://einnews.com/pr_news/593539538/ibeo-files-for-insolvency-and-holds-talks-with-potential-investors-self-administration-proceedings-have-been-granted (accessed in June 2022)
[56] https://innoviz.tech/innovizone (accessed in June 2022)
[57] https://innoviz.tech/innoviztwo (accessed in June 2022)
[58] https://techcrunch.com/2022/08/02/innoviz-lidar-will-be-on-all-vw-vehicles-with-automated-driving-capabilities/ (accessed in June 2022)
[59] https://hesaitech.com/adas/# (accessed in June 2022)
[60] https://robosense.ai/en/product (accessed in June 2022)

[61] https://just-auto.com/news/company-news/toyota-cars-to-be-powered-by-robosense-lidar/ (accessed in June 2022)
[62] https://businesswire.com/news/home/20230207005749/en/Future-Toyota-Cars-to-be-Powerd-by-RoboSense-LiDAR-for-Safer-Self-Driving (accessed in June 2022)
[63] https://iotworldtoday.com/transportation-logistics/robosense-to-supply-lidar-for-self-driving-toyotas (accessed in June 2022)
[64] https://huaweicentral.com/huawei-has-3-of-the-overall-lidar-market-report/ (accessed in June 2022)
[65] https://livox-wiki-en.readthedocs.io/ (accessed in June 2022)
[66] https://livoxtech.com/hap (accessed in June 2022)
[67] https://prweb.com/releases/xaos_motors_exhibits_mass_producible_ai_mems_lidar_at_ces_2021/prweb17657780.htm (accessed in July 2022)
[68] https://global.pioneer/en/corp/news/press/2020/pdf/1210-1.pdf (accessed in July 2022)
[69] https://mitsubishielectric.com/news/2020/0312.html (accessed in July 2022)
[70] https://news.panasonic.com/global/press/en170911-2 (accessed in July 2022)
[71] https://omron.com/global/en/media/press/2018/09/c0913.html (accessed in July 2022)
[72] https://components.omron.com/us-en/solutions/sensor/3d-tof-sensor-module (accessed in July 2022)
[73] https://konicaminolta.com/newsroom/2017/0328_01_01.html (accessed in July 2022)
[74] https://hokuyo-usa.com/products (accessed in July 2022)
[75] https://asia.toshiba.com/highlights/toshiba-takes-on-the-worlds-smallest-lidar-improving-the-eyes-of-autonomous-driving/ (accessed in July 2022)
[76] https://global.toshiba/ww/technology/corporate/rdc/rd/topics/22/2203-02.html (accessed in July 2022)
[77] https://techinsights.com/blog/sony-d-tof-sensor-found-apples-new-lidar-camera (accessed in July 2022)
[78] https://sony-semicon.com/en/products/is/automotive/tof.html (accessed in July 2022)
[79] https://baraja.com/en/spectrum-hd25 (accessed in August 2022)
[80] Zhao F, Jiang H and Liu Z 2019 Recent development of automotive LiDAR technology, industry and trends *11th Int. Conf. on Digital Image Processing (ICDIP 2019)*
[81] https://crunchbase.com/organization/neuvition (accessed in August 2022)
[82] https://crunchbase.com/organization/insight-lidar (accessed in August 2022)
[83] Insight LiDAR Inc. 2022 Multi-beam, wavelength-tuned laser scanning *US Patent Specification* 11493753B1
[84] https://optics.org/news/12/12/11 (accessed in August 2022)
[85] https://silc.com/product/ (accessed in August 2022)
[86] https://xenomatix.com/lidar/xenolidar/ (accessed in August 2022)
[87] https://secpho.org/en/noticias/beamagine-a-new-spin-off-company-from-cd6/ (accessed in August 2022)
[88] https://beamagine.com/products/ (accessed in August 2022)
[89] https://hybrid-lidar.com/gallery-en.html (accessed in August 2022)
[90] https://blickfeld.com/technology/ (accessed in August 2022)
[91] https://scantinel.com/technology/ (accessed in August 2022)
[92] https://scantinel.com/technology/ (accessed in August 2022)
[93] https://greencarcongress.com/2020/08/20200812-scantinel.html (accessed in August 2022)

[94] https://epic-assoc.com/wp-content/uploads/2022/02/4.5-Frank-Gindele_Scantinel.pdf (accessed in August 2022)
[95] https://epic-assoc.com/wp-content/uploads/2021/12/Claude-Florin-Fastree3D.pdf (accessed in October 2022)
[96] https://fastree3d.com/solutions/#product (accessed in October 2022)
[97] https://crunchbase.com/organization/opsys-tech (accessed in October 2022)
[98] https://opsys-tech.com/technology (accessed in October 2022)
[99] https://opsys-tech.com/post/opsys-tech-scanning-microflash-lidar-outshines-competitors-with-unrivaled-scan-rate (accessed in October 2022)
[100] https://opsys-tech.com/post/opsys-signed-a-collaboration-to-supply-pure-solid-state-lidar-to-hasco-automotive (accessed in October 2022)
[101] https://golden.com/wiki/SOS_Lab._Co.%2C_Ltd-MN5EDP5 (accessed in October 2022)
[102] https://soslab.co/ML (accessed in October 2022)
[103] https://thelec.net/news/articleView.html?idxno=1516 (accessed in October 2022)
[104] https://eetasia.com/hyundai-motor-group-and-soslab-to-co-develop-lidar-for-mobile-robots/ (accessed in October 2022)
[105] https://therobotreport.com/hesai-technology-announces-fully-solid-state-lidar/ (accessed in October 2022)
[106] https://lslidar.com/products/ (accessed in October 2022)
[107] https://panasonic.com/global/business/p-its/2021/on-chip_beam_steering_lidar.html (accessed in October 2022)

www.ingramcontent.com/pod-product-compliance
Ingram Content Group UK Ltd.
Pitfield, Milton Keynes, MK11 3LW, UK
UKHW051849100926
8556IPUK00005B/527

9 780750 337212